SCIENCE IN A BOX

26 Easy-to-Do Science Activities and Center Ideas

Life Science
- Seeds
- Leaves
- Flowers
- Animal Habitats
- Animal Coverings
- Life Cycles

Earth Science
- Weather
- Soil
- Rocks
- The Environment
- The Sun
- Moon Phases

Physical Science
- Solids
- Liquids
- Sound
- Motion

And Ten Other Topics!

Managing Editor: Debra Liverman

Editorial Team: Becky S. Andrews, Kimberley Bruck, Karen P. Shelton, Diane Badden, Thad H. McLaurin, Peggy W. Hambright, Hope Taylor Spencer, Karen A. Brudnak, Sarah Hamblet, Hope Rodgers, Dorothy C. McKinney, Amy Barsanti, Stacie Stone Davis, Kelli Gowdy, Liz Harrell, Karen Pavlosky, Jenice Pearson, Laura Wagner, Joyce Wilson

Production Team: Lisa K. Pitts, Pam Crane, Rebecca Saunders, Jennifer Tipton Cappoen, Chris Curry, Sarah Foreman, Theresa Lewis Goode, Clint Moore, Greg D. Rieves, Barry Slate, Donna K. Teal, Zane Williard, Tazmen Carlisle, Irene Harvley-Felder, Amy Kirtley-Hill, Kristy Parton, Cathy Edwards Simrell, Lynette Dickerson, Mark Rainey, Debbie Shoffner

www.themailbox.com

Table of Contents

How to Use *Science in a Box* 3
Reproducible Labels 4
Life Science Activities 5
Earth Science Activities 101
Physical Science Activities 143

Skills Grid

Activity Title	Skill	Page
Life Science		
Seeds	identifying the parts of a seed	5
Stems	understanding the function of a plant stem and examining different kinds of stems	11
Leaves	understanding that leaves have different characteristics	17
Flowers	identifying the parts of a flower and understanding their functions	23
Plant Life Cycle	recognizing and sequencing the stages of a plant's life cycle	29
Plant Growth	understanding how water and sunlight affect plant growth	35
Classifying Animals	classifying animals according to their movements	41
Animal Needs	understanding how different animals meet needs	47
Animal Coverings	exploring different animal coverings and their functions	53
Animal Feet	understanding how an animal's feet help it survive in its environment	59
Animal Mouths	examining how an animal's mouth helps it get food	65
Animal Wings	exploring the importance, functions, and differences of various animal wings	71
Animal Habitats	matching animals to their habitats	77
Animal Adaptations	understanding how a polar bear has adapted to its environment	83
Animal Life Cycles	comparing the life cycles of a grasshopper and a frog	89
Interdependence of Plants and Animals	understanding the interdependence of animals and plants	95
Earth Science		
Observing Weather	observing and recording changes in weather	101
Seasons	learning about how seasons change in different geographic locations	107
Soil	observing and describing different types of soil	113
Rocks	understanding that rocks can be classified by the ways they are formed	119
Environment	understanding that humans can impact the environment in both safe and harmful ways	125
Position of the Sun	understanding that the position of the sun changes in the sky as the earth rotates	131
Phases of the Moon	understanding the lunar cycle	137
Physical Science		
Solids and Liquids	understanding that a solid has a definite shape and a liquid does not	143
Sound	understanding that vibrations cause sound	149
Force and Motion	understanding that a force (a push or a pull) causes motion	155

©2005 The Mailbox® Books
All rights reserved.
ISBN10 #1-56234-657-1 • ISBN13 #978-1-56234-657-7

Except as provided for herein, no part of this publication may be reproduced or transmitted in any form or by any means, electronic or mechanical, including photocopying, recording, or storing in any information storage and retrieval system or electronic online bulletin board, without prior written permission from The Education Center, Inc. Permission is given to the original purchaser to reproduce patterns and reproducibles for individual classroom use only and not for resale or distribution. Reproduction for an entire school or school system is prohibited. Please direct written inquiries to The Education Center, Inc., P.O. Box 9753, Greensboro, NC 27429-0753. The Education Center®, *The Mailbox*®, the mailbox/post/grass logo, and The Mailbox Book Company® are registered trademarks of The Education Center, Inc. All other brand or product names are trademarks or registered trademarks of their respective companies.

HPS 227866

How to Use *Science in a Box*

1. Select an activity from the skills grid on page 2.
2. Read the corresponding teacher page and gather the needed materials.
3. Follow the directions below to assemble the storage box for the activity.
4. Do the whole-class activity with your students.
5. If desired, use the center idea provided on the teacher page for further reinforcement.

Sample Activity Pages
(Activity contents may vary.)

Teacher Page

Recording Sheet

Color Mini Poster

Center Activity Page

How to Assemble Each Box

1. Program a copy of the appropriate label on page 4 with the name of the selected activity.
2. Cut out and glue the label onto one end of a shoebox.
3. Fill the box as directed on the teacher page.

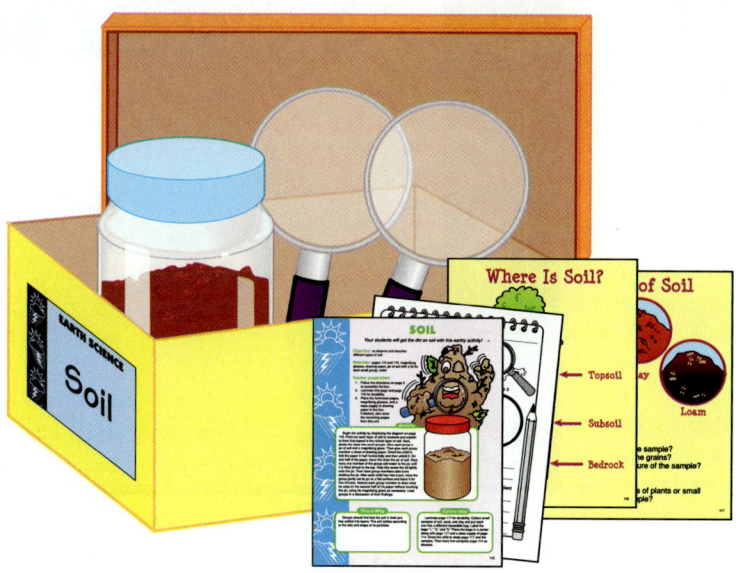

Labels
Use with "How to Assemble Each Box" on page 3.

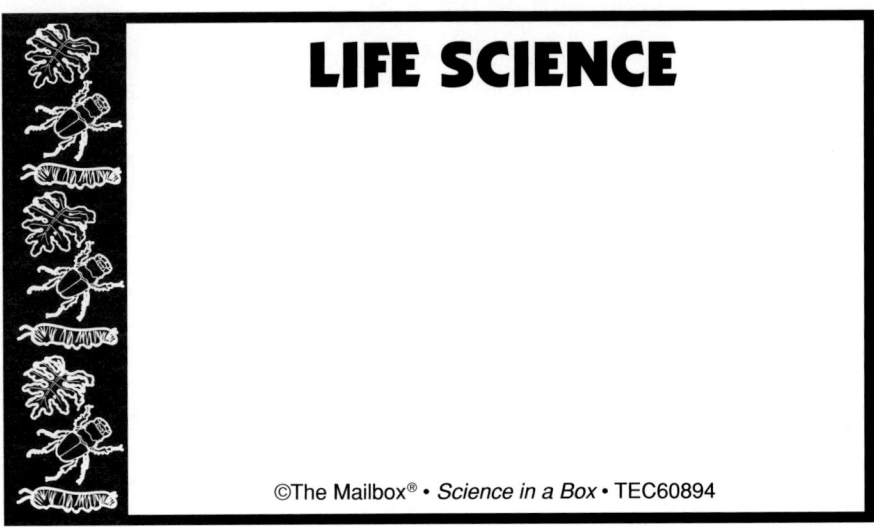

SEEDS

Objective: to identify the parts of a seed

Materials: pages 5, 7, and 9; class supply of lima beans; toothpicks; index cards

Teacher preparation:
1. Follow the directions on page 3 to assemble the box.
2. Soak the beans in water overnight.
3. Laminate this page and the seed cards on pages 7 and 9 for durability. Cut apart the cards.
4. Place this page, the seed cards, the toothpicks, and the index cards inside the box.

Activity

Begin the activity by displaying the seed cards one at a time. Invite students to comment on each seed's shape, color, and size. Explain that although seeds don't look the same, they all have the same function: to produce new plants. Point out that all seeds have three basic parts: a seed coat, a seedling, and stored food. Next, give each child one lima bean, a toothpick, and an index card. Show each child how to carefully run the toothpick around the edge of the bean so that the seed coat comes apart and slips off. Explain that the seed halves are the stored food that feeds the seedling. Then have him use his fingernails to pop the two halves of the seed apart. Point out the seedling, or tiny plant, that is revealed and have the child use his toothpick to gently remove the plant from the seed. Have him carefully tape the three parts of his seed on his index card and label each part as shown.

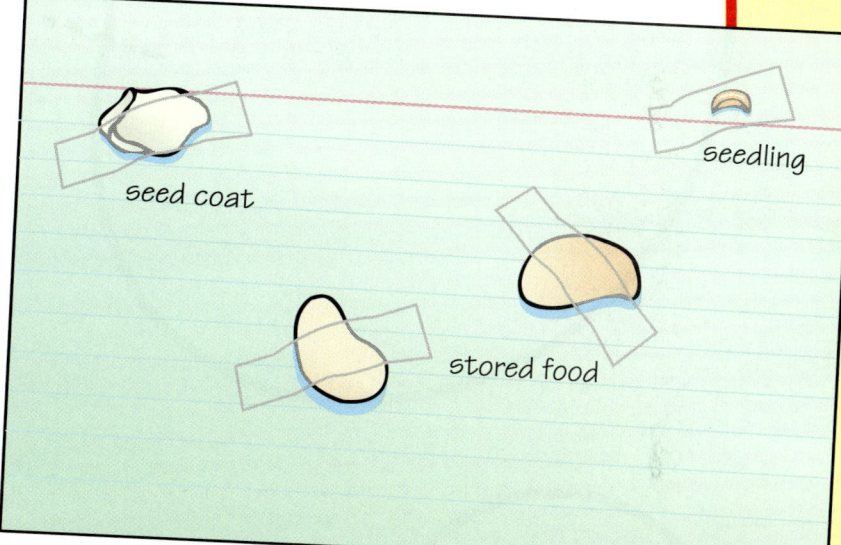

This Is Why

There are three basic parts of a seed. The *seed coat* protects the seedling that lives inside the seed. The *stored food* supplies the energy that the seedling needs to grow. The *seedling* is the part of the seed that becomes the plant.

Center Idea

Place a class supply of page 6, construction paper, crayons, scissors, and glue in a center. A student colors and cuts out the seed part patterns. Next, he glues the patterns on a sheet of construction paper, forming a seed diagram. Then he labels each seed part as shown.

©The Mailbox® • Science in a Box • TEC60894

Seed Part Patterns

Use with the center idea on page 5.

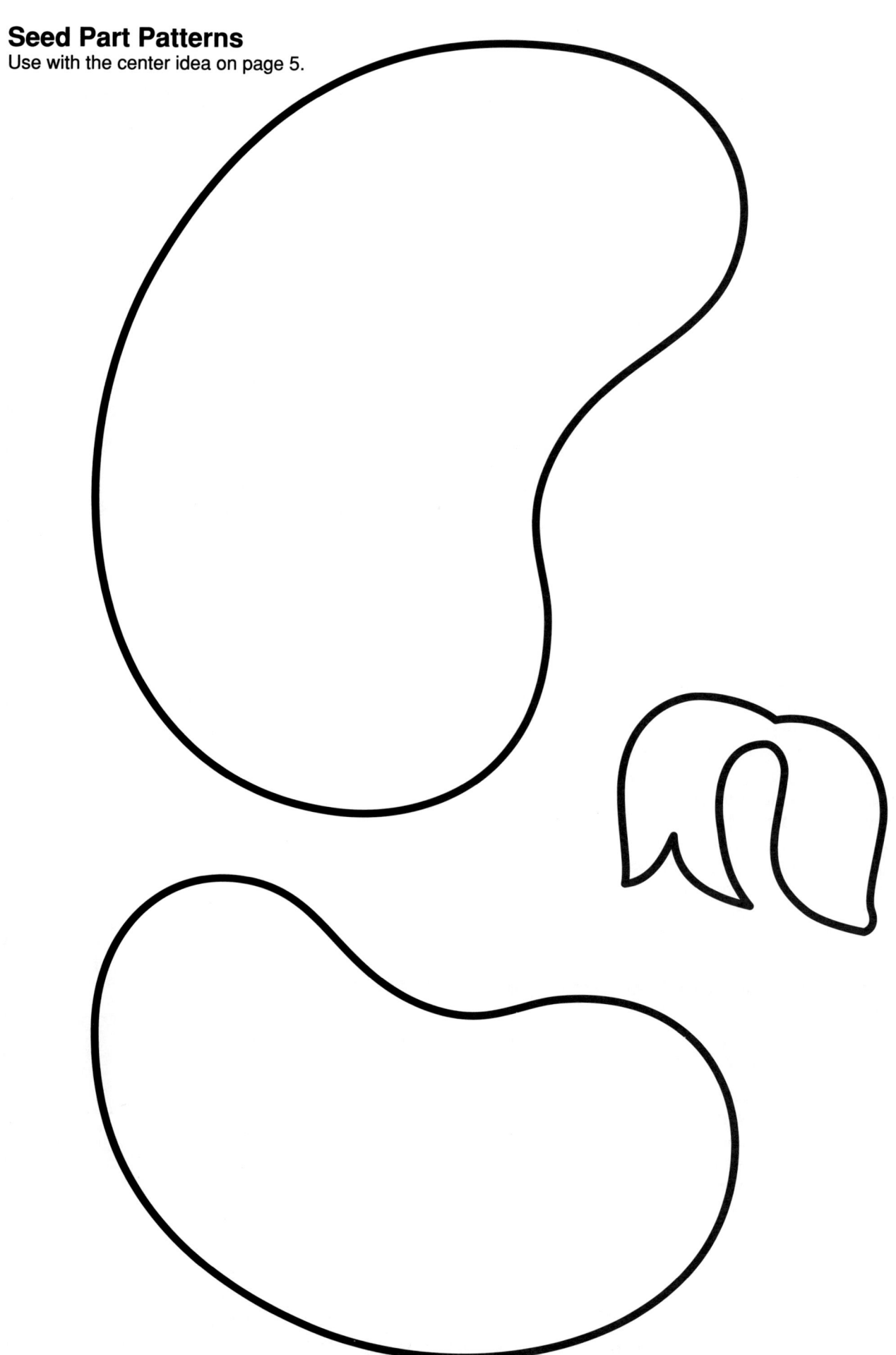

Seed Cards
Use with the activity on page 5.

Seed Cards
Use with the activity on page 5.

STEMS

Objective: to understand the function of a plant stem and examine different kinds of stems

Materials: pages 11–14, woody stem and flexible stem (optional)

Teacher preparation:
1. Follow the directions on page 3 to assemble the box.
2. Laminate this page and the stem cards on pages 13 and 14 for durability. Cut apart the cards.
3. Make a class supply of page 12.
4. Place this page, the copies, and the stem samples (optional) in the shoebox. If desired, also store the remaining pages from this unit in the box.

Activity

Begin by having students think about and describe a plant stem. After several responses, share some of the stem cards with students. Have student volunteers point out each stem in the pictures. Next, ask students to explain what they think a stem does for a plant. Share that a stem has several functions. First, it connects the leaves and roots. Second, it supports the leaves and flowers. Third, it carries water and nutrients from the roots to the leaves. Ask students whether all stems look alike. Point out that some stems are hard and woody while others are soft and flexible. If desired, pass around the sample stems that you've gathered. Then share the stem cards again, this time asking students to identify each stem as hard and woody or soft and flexible. Have students describe the characteristics of each type. Conclude that all the trees and bushes have woody stems and all the others are flexible. Follow up by having each student complete a copy of page 12.

Center Idea

Place in a center scissors, glue, and a class supply of page 15 and the top half of page 16. A child cuts out the picture cards on page 16 and glues them in the correct boxes to make a stems booklet.

This Is Why

The stem is the part of a plant that supports the plant above ground and connects the roots with the leaves. Some stems are hard and woody, like those of trees, bushes, and shrubs. Other stems are soft and flexible, like those of many flowers and vegetable plants.

©The Mailbox® • Science in a Box • TEC60894

Stems

Name _____

Read the statement.
Color the correct box.

Yes　No

☐　☐　1. A stem carries water and nutrients from the roots to the leaves.

☐　☐　2. A stem holds the plant in the ground.

☐　☐　3. A stem supports a plant's leaves and flowers.

☐　☐　4. A stem can be hard and woody.

☐　☐　5. A stem makes seeds for the plant.

☐　☐　6. A stem connects the leaves and the roots.

☐　☐　7. A stem can be soft and flexible.

☐　☐　8. A stem uses water, light, and air to make food for the plant.

Woody Stem Cards
Use with the activity on page 11.

Flexible Stem Cards
Use with the activity on page 11.

Booklet Cover and Pages
Use with the center idea on page 11.

Booklet Page and Picture Cards

Use with the center idea on page 11.

Answer Key

Page 12
1. yes
2. no
3. yes
4. yes
5. no
6. yes
7. yes
8. no

Booklet
Page 1: leaves, roots
Page 2: water
Page 3: tree, bush
Page 4: daisy, tomato plant

LEAVES

Objective: to understand that leaves have different characteristics

Materials: pages 17, 19, and 21

Teacher preparation:
1. Follow the directions on page 3 to assemble the box.
2. Laminate this page and the leaf cards on pages 19 and 21 for durability.
3. Cut apart the leaf cards.
4. Place this page and the leaf cards inside the box.

Activity

Begin the activity by explaining to students that although the leaves on different plants may be very different in appearance, each leaf performs the same function for most plants: making food. Ask students to describe a leaf. List responses on the board. Ask students to study the descriptions they've generated and then suggest categories that describe the ways that leaves are different. For example, students may say that leaves can be different colors, shapes, and sizes. Continue the discussion by explaining to students that a leaf can be identified by the shape of its edge. Explain that the broad-leaf group has three edge shapes: smooth, toothed, and lobed. Write the three edge types across the top of the board; then post the mountain laurel (smooth), holly (toothed), and maple leaf (lobed) cards underneath the appropriate headings. Have students discuss the differences in the three leaves posted on the board. Next, pass out the remaining leaf cards to 12 students. Have each student study his leaf, get input from his classmates, and then tape the leaf below the heading under which he thinks his leaf belongs. As a class, check the leaves in each category and make any needed changes.

This Is Why

The shape of the leaves varies from plant to plant. Toothed leaves help the plant get rid of extra water. In some plants, young toothed leaves release a liquid that protects the plant from insects. The lobed leaves of some plants help heat escape from the leaf. Smooth-edged leaves are common on plants found in warm climates.

Center Idea

Place the leaf cards at a center along with a class supply of page 18. A student selects two leaf cards and illustrates each one in the space provided on page 18. Then he completes the recording sheet as directed.

©The Mailbox® • Science in a Box • TEC60894

How Are Leaves Different?

Name _____

[leaf image box]
_____
name of leaf

[leaf image box]
_____
name of leaf

1. Write two ways the leaves are alike. _____
_____
_____
_____

2. Write two ways the leaves are different. _____
_____
_____
_____

Leaf Cards
Use with the activity and center idea on page 17.

Leaf Cards
Use with the activity and center idea on page 17.

FLOWERS

Objective: to identify the parts of a flower and understand their functions

Materials: pages 23–27, several different silk (or fresh) flowers with easily identifiable flower parts (optional)

Teacher preparation:
1. Follow the directions on page 3 to assemble the box.
2. Laminate this page and the flower cards (if not using silk flowers) on page 25 for durability. Cut apart the cards.
3. Make a class supply of page 27.
4. Place this page, the flower cards (or the silk flowers) and the copies in the box. If desired, also store the remaining page from this unit in the box.

Activity

Begin by holding up the flower cards (or silk flowers) for students to examine. Have students point out the similarities among the different flowers. Tell students that flowers come in many different colors and sizes, but they all make seeds for plants. Next, distribute copies of page 27 and display the color diagram from page 24. Share the different parts of a flower and their functions. Then have students fill in their diagrams to match the information on your color version. Finally, have students reexamine the flower cards (or silk flowers) and identify the four flower parts on each.

This Is Why

The flower is the part of a plant that produces seeds. The *sepals* protect the inner parts of the flower as it grows. The *pistil* is where the seeds develop and grow. The *stamen* produces a powdery material called pollen that is used in pollination. *Petals* are the brightly colored part of most flowers. They attract insects and birds that help spread the pollen.

Center Idea

Place a class supply of page 28 and crayons at a center. If desired, also place the color diagram from page 24 at the center for students to use as a guide. A student visiting the center completes page 28, referring to the diagram as needed.

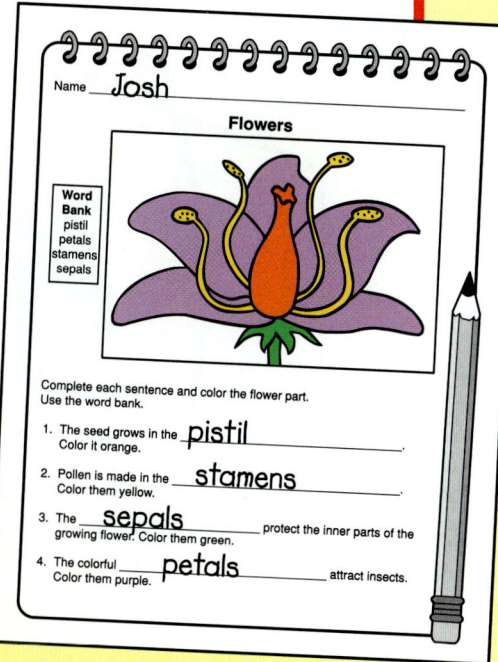

©The Mailbox® • *Science in a Box* • TEC60894   23

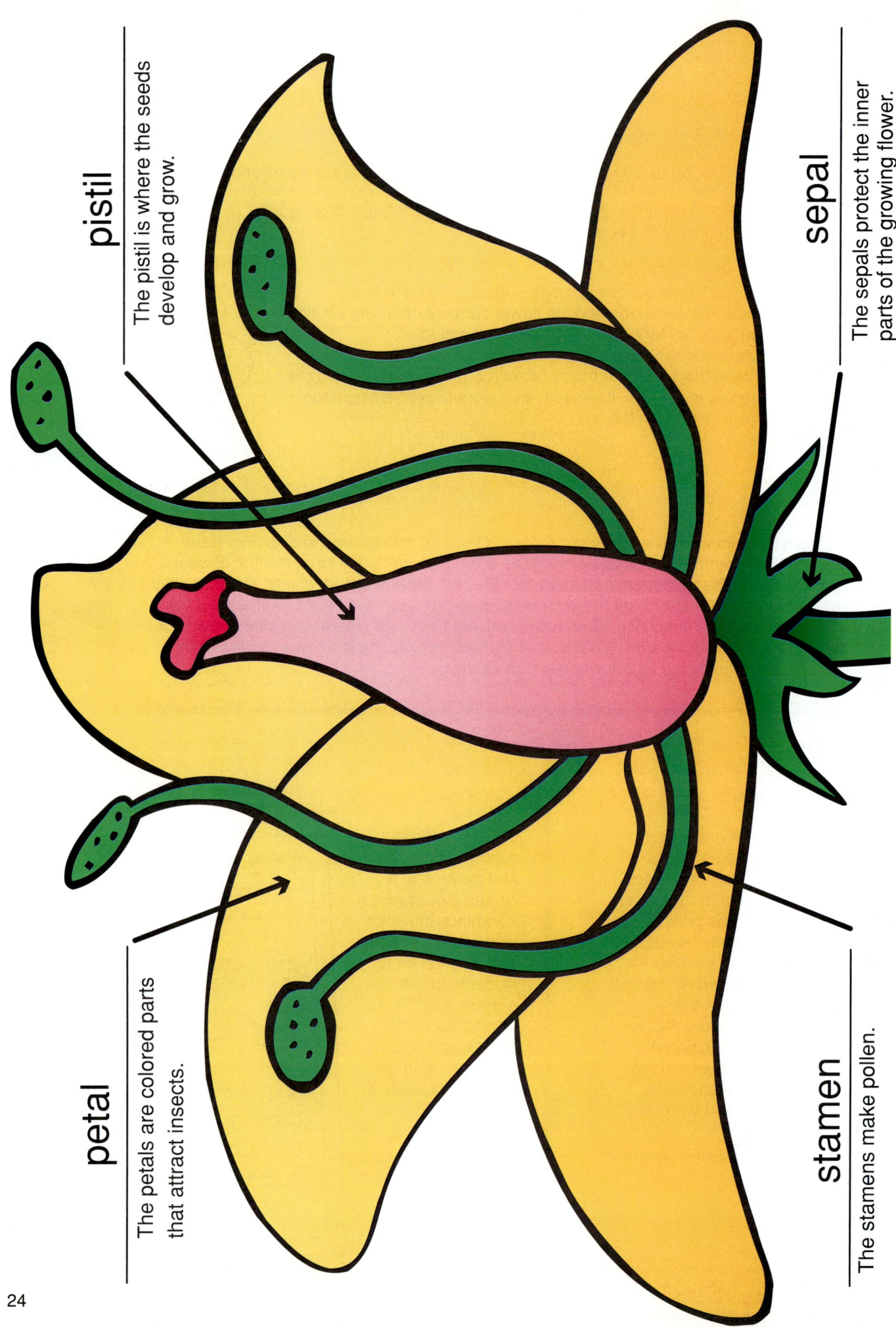

Flower Cards
Use with the activity on page 23.

25

Name _____

Note to the teacher: Use with the activity on page 23.

Name _____

Flowers

Word Bank
pistil
petals
stamens
sepals

Complete each sentence and color the flower part. Use the word bank.

1. The seed grows in the _____.
   Color it orange.

2. Pollen is made in the _____.
   Color them yellow.

3. The _____ protect the inner parts of the growing flower. Color them green.

4. The colorful _____ attract insects.
   Color them purple.

PLANT LIFE CYCLE

Objective: to recognize and sequence the stages of a plant's life cycle

Materials: pages 29, 31, and 33

Teacher preparation:
1. Follow the directions on page 3 to assemble the box.
2. Laminate this page and the plant life cycle cards on pages 31 and 33 for durability. Cut apart the cards.
3. Place this page and the life cycle cards in the box.

Activity

Begin by displaying the plant life cycle card that reads "A seed absorbs water and starts to swell." Lead students to discuss the card and then invite several students to predict what happens next in the life cycle. Continue in this manner, showing each of the remaining cards in order. Point out that these stages always take place in the same order.

Next, give one card to each of six students. Direct the students with cards to arrange themselves in correct order according to the life cycle. If students line up in a straight line, challenge them to think of a way to show that the cards describe a cycle *(standing in a circle)*. Collect, shuffle, and redistribute the cards to six different students. Repeat until each child has had a turn to hold a card.

Center Idea

Place the life cycle cards, a class supply of page 30, sheets of construction paper, crayons, scissors, and glue at a center. A student uses the word bank to complete each sentence, referring to the life cycle cards as needed. Then he colors and cuts out the petals. Next, he glues the petals in order on the construction paper, forming a flower as shown.

This Is Why

The stages of a flowering plant's life cycle are as follows: (1) the seed germinates, (2) the seed grows into a new plant, (3) the plant matures, (4) the flowers open and the plant is pollinated, (5) the flowers wither and seeds develop, and (6) the seeds are dispersed.

©The Mailbox® • *Science in a Box* • TEC60894

Life Cycle Petals
Use with the center idea on page 29.

Word Bank
flowers   Leaves
Buds   stem
roots   Seeds
seed

1. A _____ absorbs water and starts to swell.

2. The _____ begin to grow. Then the _____ appears.

3. _____ begin to develop on the stem.

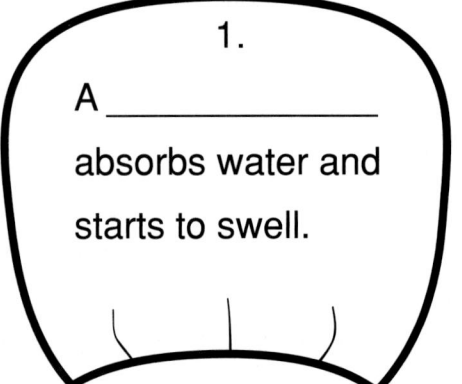

4. The plant grows. _____ begin to show up.

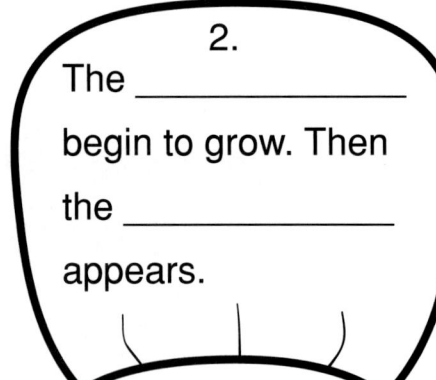

5. The _____ open. They are pollinated.

6. The flowers wilt. _____ develop. The seeds are spread around.

Plant Life Cycle Cards
Use with the activity and center idea on page 29.

Plant Life Cycle Cards

Use with the activity and center idea on page 29.

PLANT GROWTH

Objective: to understand how water and sunlight affect plant growth

Materials: pages 35 and 36, seedlings planted in 6 plastic cups or small pots

Teacher preparation:
1. Follow the directions on page 3 to assemble the box.
2. Laminate this page for durability.
3. Make a class supply of page 36.
4. Place this page and the copies in the box. If desired, also store the remaining pages from this unit in the box.

Activity

Begin by having students brainstorm ideas about what might happen to a plant if it receives too much or too little water and sunlight. Have each child record his predictions in the chart at the top of page 36. Afterward, designate three plants for the water test and three for the sunlight test. Label the first group of plants A–C. Label the second group of plants C–E. Explain that the plant labeled "C" in each group will be the control plant. Next, guide students to design a fair investigation to test their predictions. Conduct the experiment over a period of two to three weeks. Then have students make and record their observations in the chart at the bottom of page 36.

This Is Why

Plants need water, sunlight, the right temperature, and good soil to grow. The amount of water and sunlight a plant receives can affect its size, the rate at which it grows, and whether it lives or dies.

Center Idea

Laminate the gameboard and cards on pages 37 and 39 for durability. Cut apart the cards and place them at a center along with the gameboard and four dried beans to use as game markers. Invite two to four players to visit the center and play the game as directed.

©The Mailbox® • Science in a Box • TEC60894

35

Name _____

My Predictions

Date: _____

Factor	Amount	What Will Happen
Water	too much	
	too little	
Sunlight	too much	
	too little	

My Observations

Date: _____

Plant	Water/Sunlight	Amount	What I Observed
A			
B			
D			
E			

©The Mailbox® • Science in a Box • TEC60894

36 **Note to the teacher:** Use with the activity on page 35.

Growth Spurt

Directions:
1. Place the cards facedown on the board.
2. Place your game pieces on Start.
3. Choose a card and follow the directions.
4. The first player to Finish wins.

Place game cards here.

Start

Finish

©The Mailbox® • *Science in a Box* • TEC60894

37

Game Cards

Use with the center idea on page 35.

Flooding! Move back 1 space.	Insects are eating your leaves! Move back 1 space.	Fertilizer! Move ahead 1 space and take another turn.	Tender loving care! Move ahead 4 spaces.	Earthworms are moving through the soil! Move ahead 1 space.
Sunshine! Move ahead 3 spaces.	Gentle rain! Move ahead 2 spaces.	It's freezing! Lose a turn.	Tender loving care! Move ahead 4 spaces.	Litterbug nearby! Move back 2 spaces.
Sunshine! Move ahead 3 spaces.	Gentle rain! Move ahead 2 spaces.	A house shades you in the heat of the day! Move ahead 1 space.	Too much sun! Move back 1 space.	Insects are pollinating! Move ahead 2 spaces.
Sunshine! Move ahead 3 spaces.	Gentle rain! Move ahead 2 spaces.	Weeds are gone! Move ahead 3 spaces.	Not enough water! Move back 1 space.	You're the biggest plant! Move any player back 2 spaces.

©The Mailbox® • Science in a Box • TEC60894

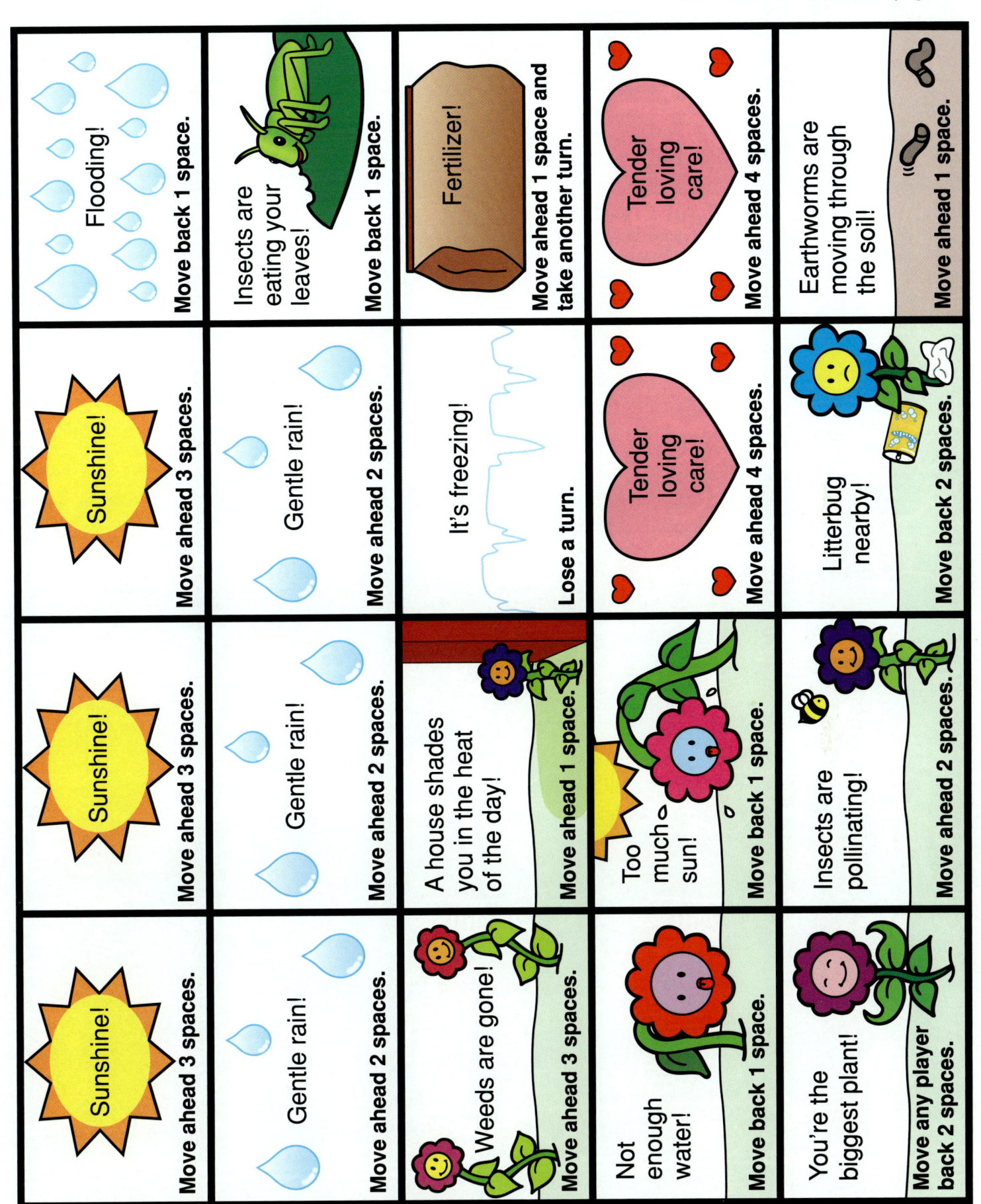

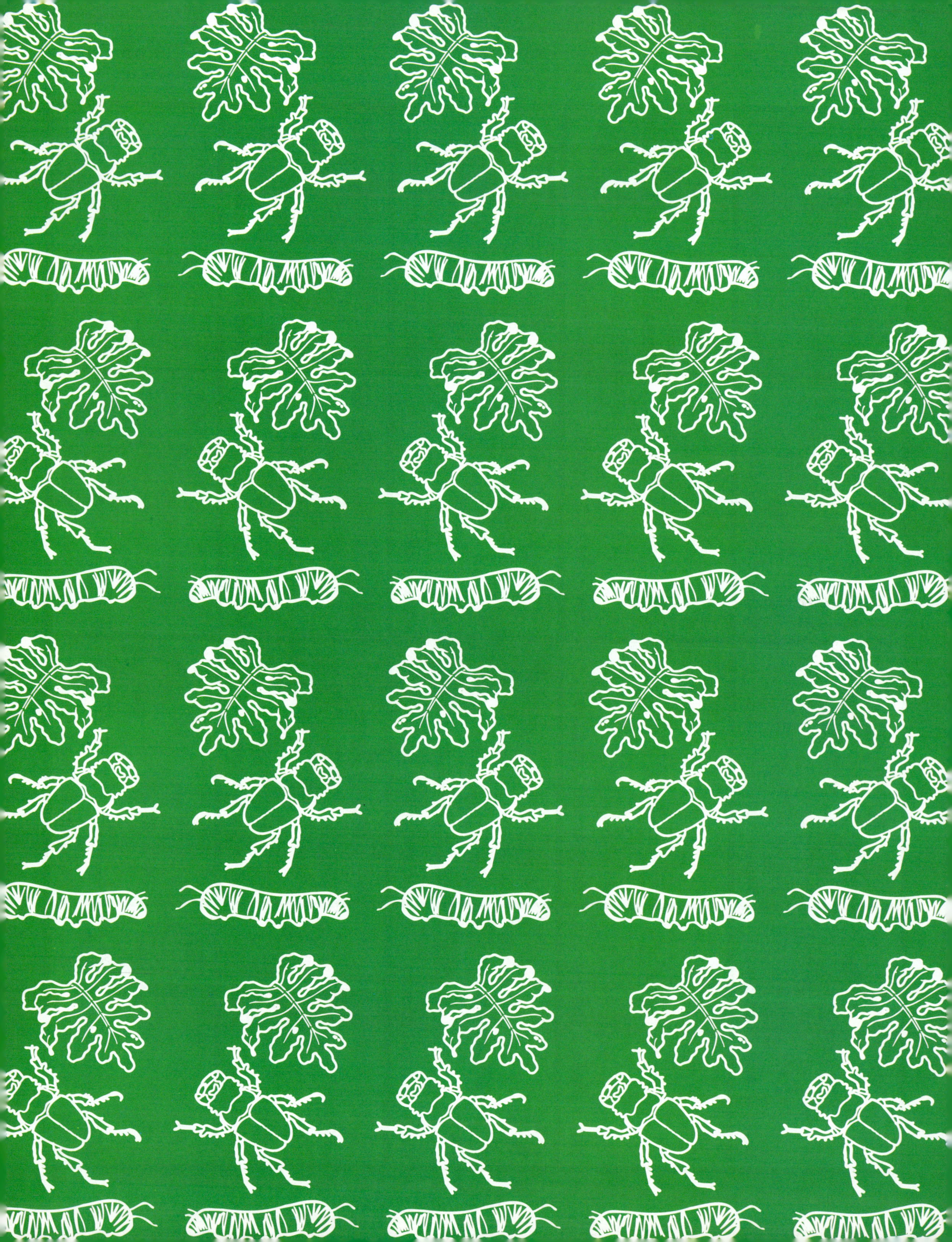

CLASSIFYING ANIMALS

Objective: to classify animals according to their movements

Materials: pages 41–44, blank transparency

Teacher preparation:
1. Follow the directions on page 3 to assemble the box.
2. Make a transparency of the gameboard on page 42. Color each game marker on the transparency a different color and then cut them out.
3. Make a copy of the animal cards on pages 43 and 44 and cut them apart.
4. Laminate this page for durability.
5. Place the gameboard transparency, game markers, and the animal cards inside the box. If desired, also store the remaining pages from this unit in the box.

Activity

Begin by taping the following animal cards to the board: butterfly, whale, rattlesnake, lion, and frog. Ask students to pretend that the animals shown are really hungry and want to get to a food market. Have students tell how each animal would move to get there. Write each movement above the appropriate card as shown. Explain that one way animals can be classified, or grouped, is by how they move. Next, stack all the cards facedown in a pile and project the gameboard transparency. Divide your class into two teams and place the game markers on Start. Have the first player from Team 1 draw a card, identify the animal, and tell whether it moves by running or walking, hopping, crawling, swimming, or flying. If correct, move that team's game marker to the next matching space. If incorrect, give the first player from Team 2 a chance to answer. Continue play in this manner. The first team to reach Checkout wins. Conclude by having the teams tape all of the game cards under the appropriate headings listed on the board.

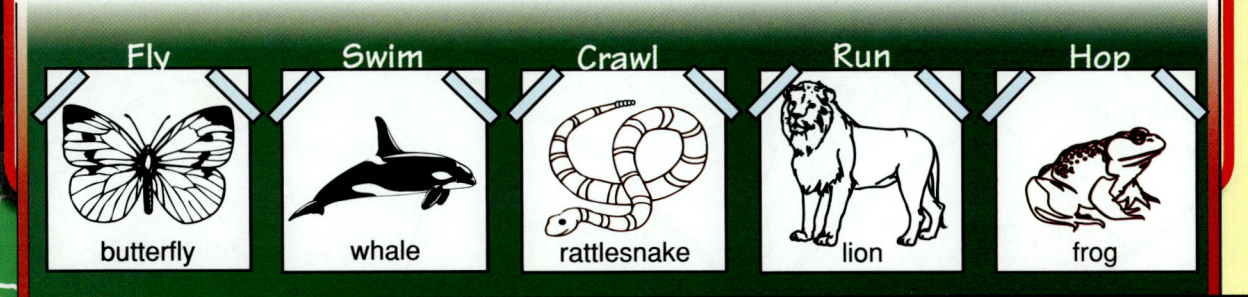

Fly	Swim	Crawl	Run	Hop
butterfly	whale	rattlesnake	lion	frog

This Is Why

One way to classify animals is to group them by how they move. Some swim or fly; others may hop, run, or crawl. Although some animals can move in more than one way, they may have a primary way in which they move.

Center Idea

Make several copies of pages 43 and 44. Cut out the cards and place them at a center along with a class supply of pages 45 and 46, scissors, and glue. A student follows the directions to complete and assemble his booklet.

©The Mailbox® • *Science in a Box* • TEC60894

To Market!

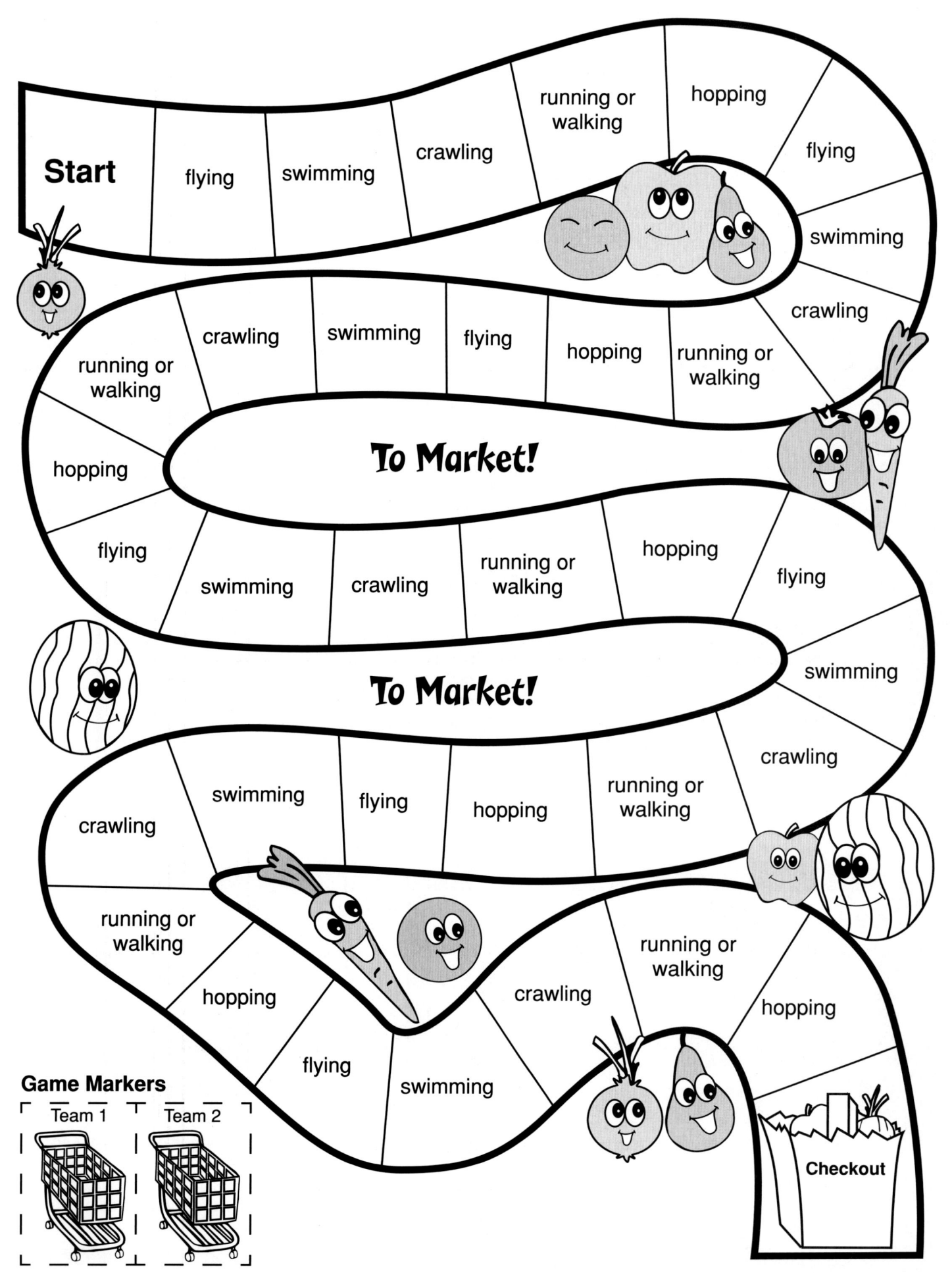

Start

flying • swimming • crawling • running or walking • hopping • flying • swimming • crawling • running or walking • hopping • flying • swimming • crawling • running or walking • hopping • crawling • swimming • flying

To Market!

swimming • crawling • running or walking • hopping • flying • swimming • crawling

To Market!

crawling • swimming • flying • hopping • running or walking • crawling • running or walking • hopping • flying • swimming • crawling • running or walking • hopping

Checkout

Game Markers
Team 1   Team 2

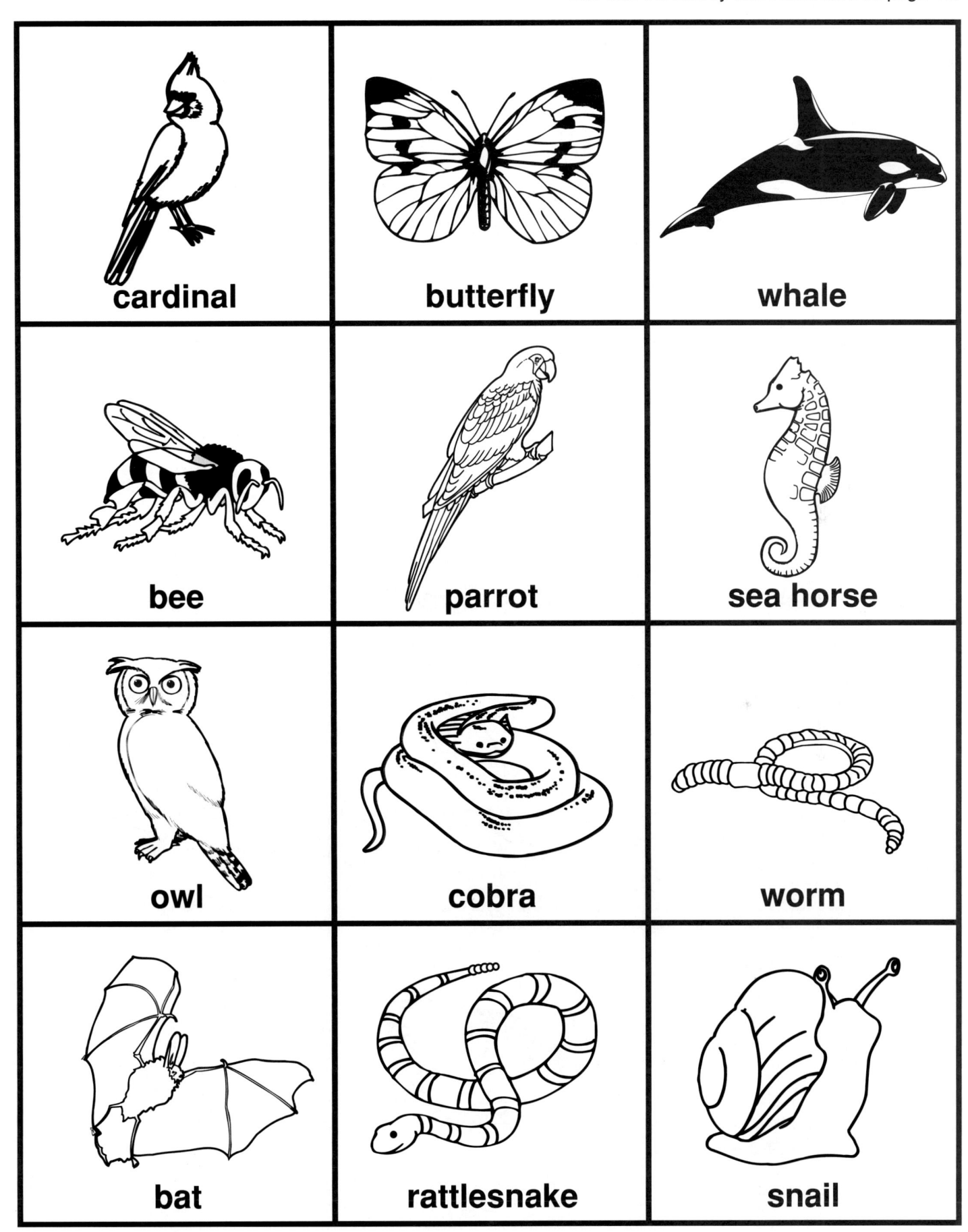

Animal Cards
Use with the activity and center idea on page 41.

cardinal	butterfly	whale
bee	parrot	sea horse
owl	cobra	worm
bat	rattlesnake	snail

©The Mailbox® • Science in a Box • TEC60894

Animal Cards
Use with the activity and center idea on page 41.

fish	**sheep**	**elephant**
eel	**grasshopper**	**lion**
shark	**kangaroo**	**mouse**
rabbit	**frog**	**ostrich**

Booklet Cover and Pages
Use with the center idea on page 41.

1

Running or Walking Animals

Glue.

3

Flying Animals

Glue.

Animal Movement

by _____

©The Mailbox® • Science in a Box • TEC60894

2

Crawling Animals

Glue.

45

Booklet Pages

Use with the center activity on page 41.

5

Swimming Animals

Glue.

Directions:
1. Cut out the pages.
2. For pages 1–5, glue an animal picture in the first box. Draw another animal that moves the same way in the second box.
3. Complete page 6.
4. Stack the pages in order. Staple.

4

Hopping Animals

Glue.

6

A(n) _____ is my favorite animal.

It moves by _____

Here is a picture of it.

ANIMAL NEEDS

Objective: to understand how different animals meet their needs

Materials: pages 47–49

Teacher preparation:
1. Follow the directions on page 3 to assemble the box.
2. Laminate this page and the animal cards on page 49 for durability. Cut apart the cards.
3. Make a class supply of page 48.
4. Place this page, the animal cards, and the copies in the box. If desired, store the remaining pages from this unit in the box.

Need:
- air
- water
- food
- shelter

Activity

Begin the activity by explaining to students that although there are many different types of animals, each one has the same basic four needs. Next, display one of the animal cards. Ask students to name the things the animal needs as you list their ideas on the board. Repeat with each of the remaining cards. Then draw students' attention to the similar items in the lists, concluding that each animal needs air, water, food, and shelter. Point out that animals meet their needs in different ways. For example, a goldfish gets its oxygen from water, gets water from its surroundings, eats plants for food, and uses the water as its shelter. Compare this with another animal, such as a bear. A bear breathes air, drinks water, eats plants and hunts animals, and may use a den for shelter. Discuss how each of the six animals meets its four basic needs. Then have each child complete a copy of page 48.

dog
breathes air
drinks water
eats dog food or scraps
may live in a doghouse or
   owner's house

bird
breathes air
drinks water
eats worms and
   seeds
may live in a nest

turtle
breathes air
drinks water
eats plants and animals
uses its shell for shelter

This Is Why

Every animal has four needs: food, water, shelter, and air. Animals that live on land get oxygen from the air. Many water animals get their oxygen from the water. Animals drink the water they need or get it from the foods they eat. They get food by eating other animals or plants. Animals obtain shelter in many ways, such as by building nests, digging tunnels, or crawling into hollow logs.

Center Idea

Laminate the directions on page 49 and the gameboards on page 51. Place the gameboards and directions in a center along with a paper clip, a pencil, and a supply of game markers or Unifix cubes. Direct students to visit the center in pairs and follow the directions to play the game.

©The Mailbox® • Science in a Box • TEC60894

Name _____

Home, Sweet Home

Write the name of an animal. _____
Draw the animal.

What four things does your animal need to live?

_____   _____
_____   _____

Draw a place for your animal to live.
Remember to show the four things the animal needs.

©The Mailbox® • Science in a Box • TEC60894

Note to the teacher: Use with the activity on page 47.

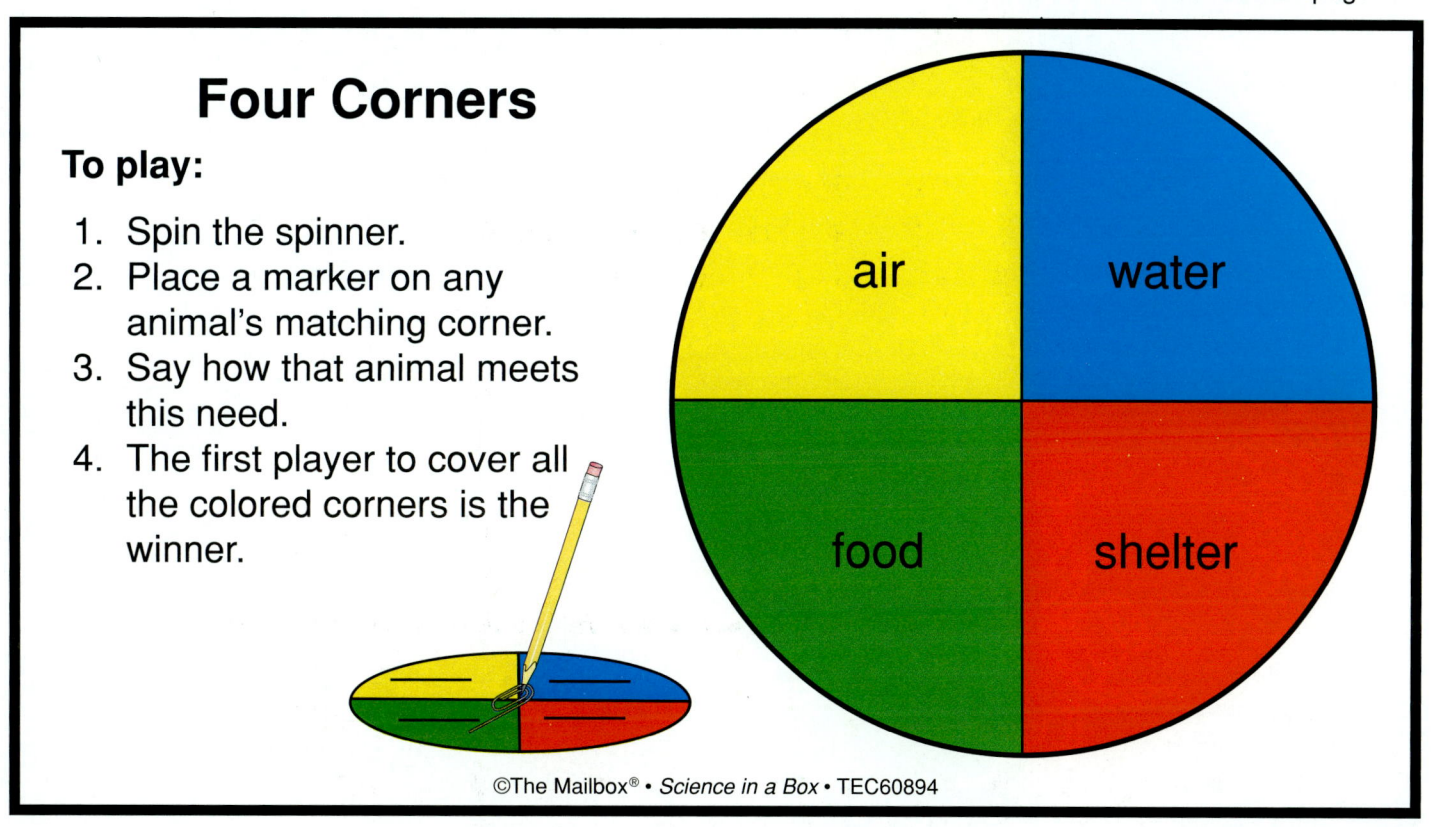

Center Directions
Use with the center idea on page 47.

Four Corners

To play:
1. Spin the spinner.
2. Place a marker on any animal's matching corner.
3. Say how that animal meets this need.
4. The first player to cover all the colored corners is the winner.

- air
- water
- food
- shelter

©The Mailbox • Science in a Box • TEC60894

Animal Cards
Use with the activity on page 47.

| dog | mouse | bear |
| goldfish | bird | turtle |

©The Mailbox • Science in a Box • TEC60894

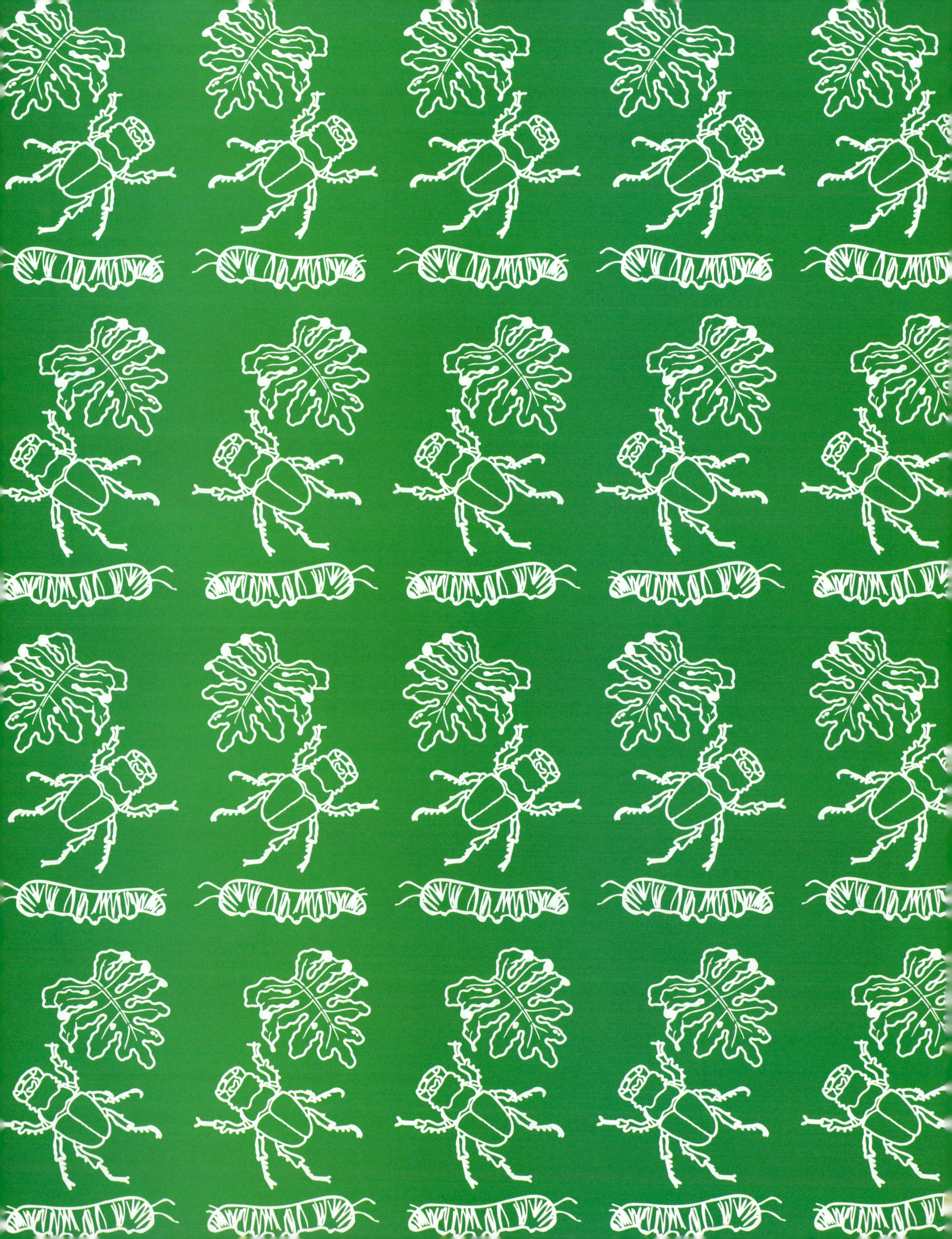

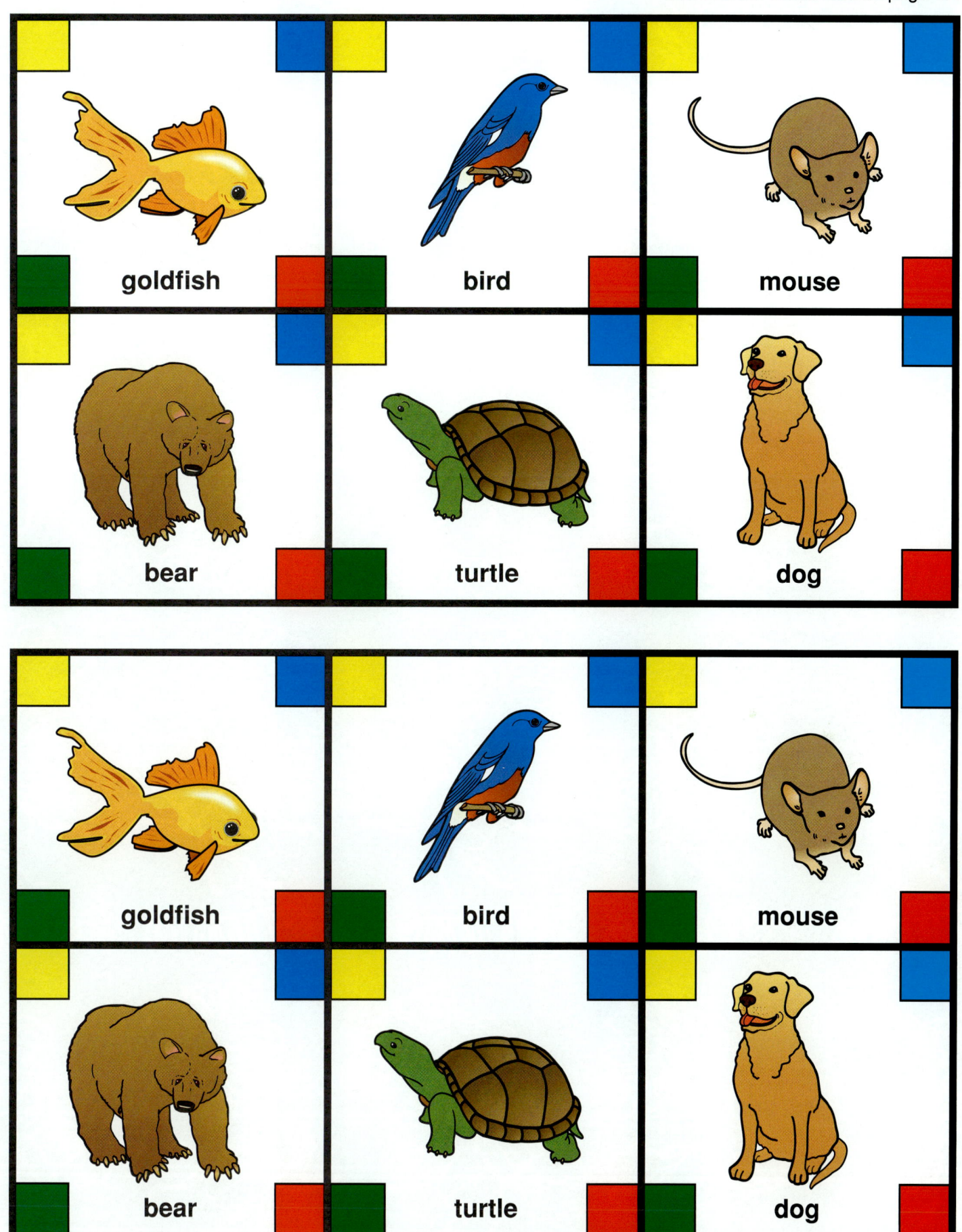

Gameboards
Use with the center idea on page 47.

goldfish	bird	mouse
bear	turtle	dog

goldfish	bird	mouse
bear	turtle	dog

©The Mailbox® • Science in a Box • TEC60894

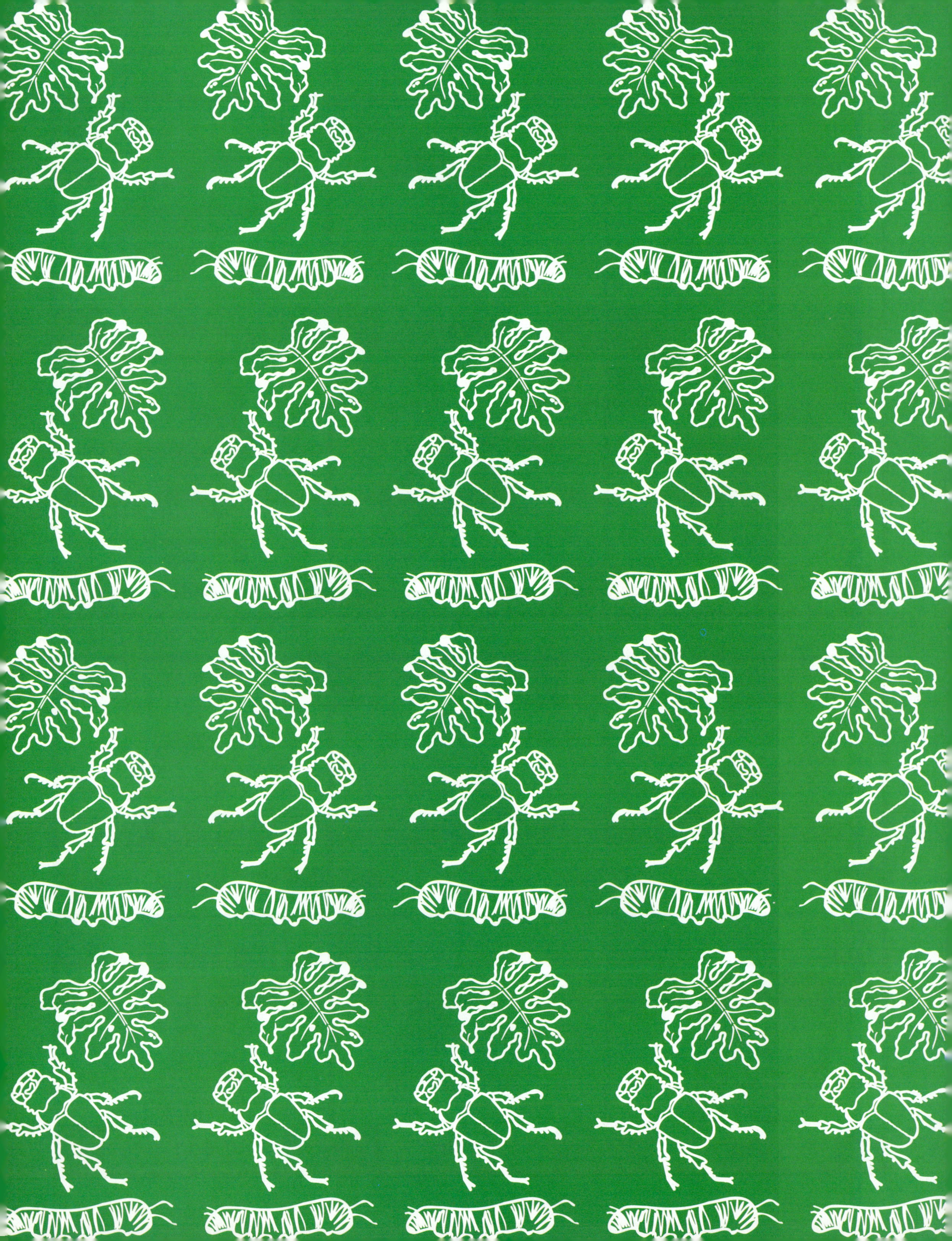

ANIMAL COVERINGS

Objective: to explore different animal coverings and their functions

Materials: pages 53–58, brad for each student, resealable bag for each small group of students

Teacher preparation:
1. Follow the directions on page 3 to assemble the box.
2. Make one copy of the animal cards on pages 57 and 58 for each small group of students. Cut apart the cards and place each set in a resealable bag.
3. Make one copy of the animal wheel patterns on pages 55 and 56 and one copy of one set of the fact strips on page 54 for each student.
4. Laminate this page for durability.
5. Place this page, the zippered bags, and the copies inside the box.

Activity

Give each group of students a plastic bag containing the animal cards. Challenge each group to sort its cards into three groups based on the animals' body coverings. After about ten minutes, ask a few groups to explain their groupings. Point out that the bear, horse, jaguar, and mouse are all covered in fur or hair. The ostrich, snowy owl, toucan, and robin are all covered in feathers. And the crocodile, snake, iguana, and fish are all covered in scales. Explain that each type of body covering is an adaptation that helps the animal survive. Then share the information from the "This Is Why" section below. Afterward, examine a few of the animals from the cards to discuss how their body coverings might help them survive.

Next, give each child a copy of pages 55 and 56, one set of fact strips from page 54, and a brad. A student cuts out the fact strips and glues them to the matching spaces on the animal wheel from page 55. Then he cuts out both wheels and uses the brad to assemble them as shown.

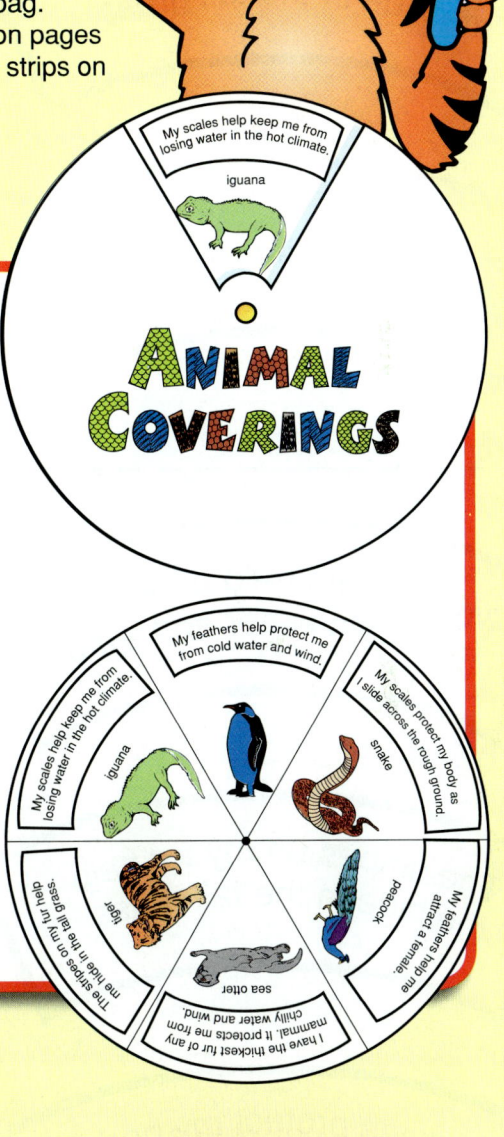

This Is Why

Feathers help birds fly and help them maintain body temperature. Some feathers help birds attract a mate or hide from enemies. Mammals have fur or hair to help keep them warm. Some mammals have fur that allows them to hide from enemies and prey. Many fish and reptiles are covered in scales, which protect a softer body underneath. Scales on reptiles help prevent water loss.

Center Idea

Make a class supply of the recording sheet at the bottom of page 58. Place the sheets and the animal cards in a zippered bag at a center. A student selects one animal card and completes the page as directed.

©The Mailbox® • Science in a Box • TEC60894

Fact Strips

Use with the activity on page 53.

- My feathers help protect me from cold water and wind.
- My feathers help me attract a female.
- The stripes on my fur help me hide in the tall grass.
- I have the thickest fur of any mammal. It protects me from chilly water and wind.
- My scales protect my body as I slide across the rough ground.
- My scales help keep me from losing water in the hot climate.

- -

- My feathers help protect me from cold water and wind.
- My feathers help me attract a female.
- The stripes on my fur help me hide in the tall grass.
- I have the thickest fur of any mammal. It protects me from chilly water and wind.
- My scales protect my body as I slide across the rough ground.
- My scales help keep me from losing water in the hot climate.

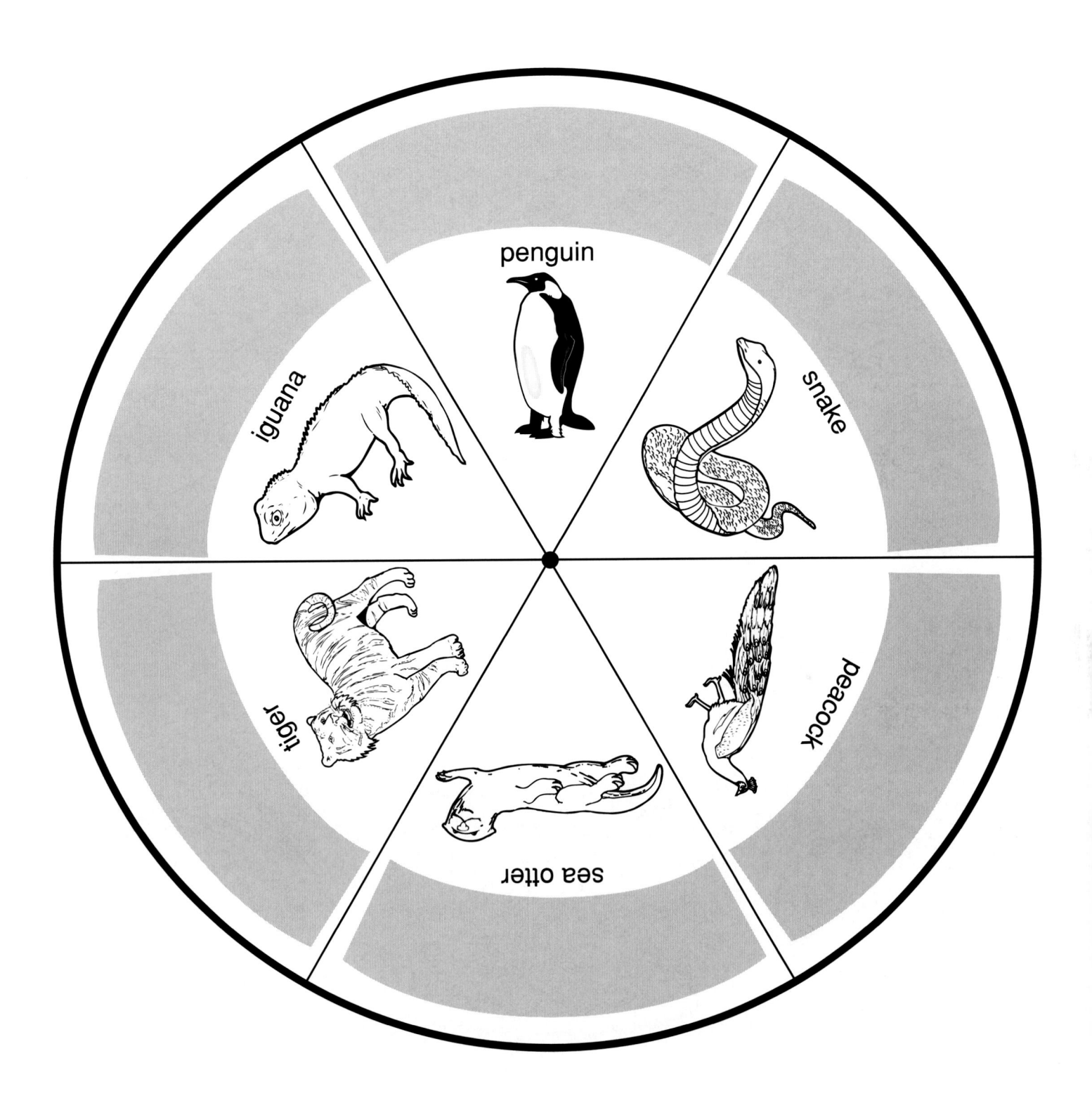

Animal Wheel Pattern
Use with the activity on page 53.

- penguin
- snake
- peacock
- sea otter
- tiger
- iguana

©The Mailbox® • Science in a Box • TEC60894

Animal Wheel Pattern
Use with the activity on page 53.

Animal Coverings

by _____

©The Mailbox® • Science in a Box • TEC60894

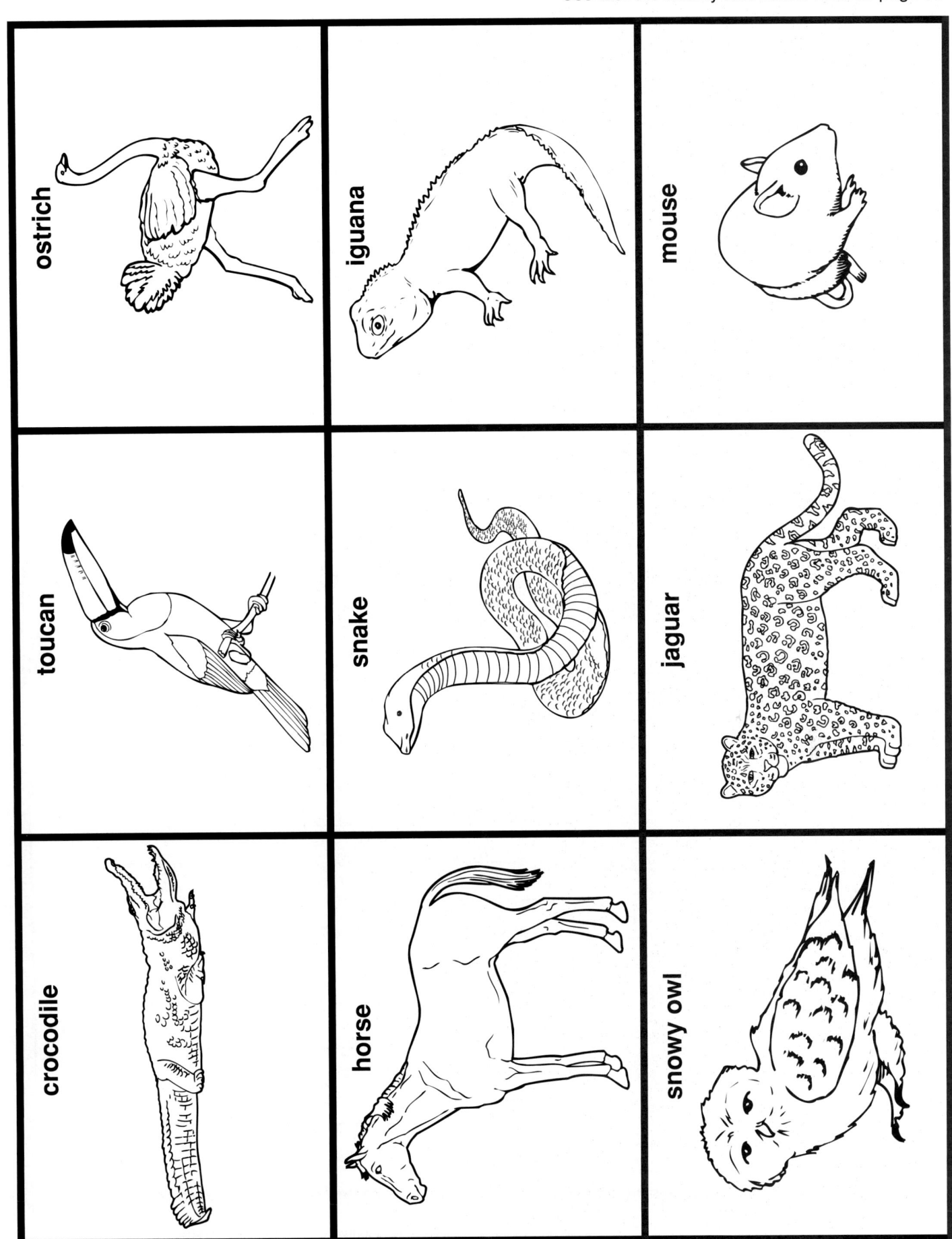

Animal Cards

Use with the activity and center idea on page 53.

ostrich	iguana	mouse
toucan	snake	jaguar
crocodile	horse	snowy owl

Animal Cards
Use with the activity and center idea on page 53.

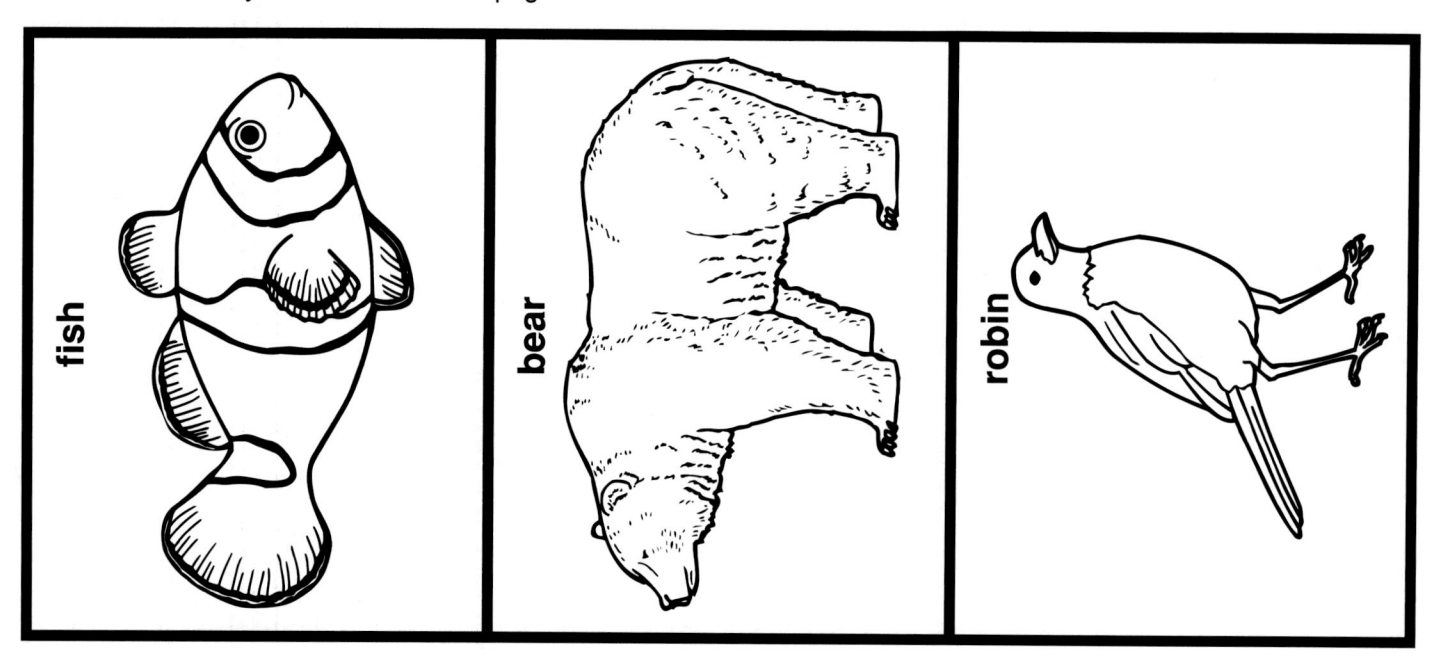

fish

bear

robin

Animal Coverings

Name _____

Select one card.
Describe the animal's covering.

_____
_____
_____

How do you think this covering helps the animal?

_____
_____
_____
_____

Draw a picture of the animal.

©The Mailbox® • Science in a Box • TEC60894

Note to the teacher: Use with the center idea on page 53.

ANIMAL FEET

Objective: to understand how an animal's feet help it survive in its environment

Materials: pages 59 and 61–64

Teacher preparation:
1. Follow the directions on page 3 to assemble the box.
2. Laminate this page and the animal cards on pages 61 and 63 for durability. Cut apart the cards.
3. Place this page and the animal cards inside the box.

Activity

To begin, display the animal cards. Explain to students that each card shows a close-up of a different animal's feet. Next, invite a pair of student volunteers to sort the cards as they see fit. To prompt students' thinking, ask questions such as the following: Where would each animal live? How does each animal move? How might its feet help the animal gather food? Have the pair share its explanation of the rules it used to classify the feet. Next, display the cards one at a time, pointing out the characteristics of each set of feet. Refer to the back of the card to explain the function of each animal's feet. Guide students to understand that each animal's feet are suited to help the animal survive in the environment in which it lives.

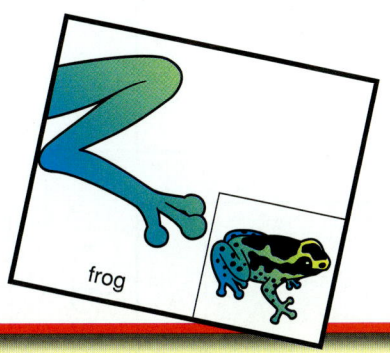

This Is Why

Animals have different kinds of feet. Some are hard, soft, curved, or pointed. Others have claws or are webbed. Animals' feet have different purposes. Some feet dig and some climb; others hop or swim. Animals may use their feet to gather food. An animal's feet are suited for survival in its environment.

Center Idea

Place the animal cards at a center along with a class supply of page 60. A student chooses three animal cards. He illustrates each foot in a space on page 60. Then he completes the page as directed.

©The Mailbox • *Science in a Box* • TEC60894

Name _____

Animal Feet

| 1. _____ Animal | 2. _____ Animal | 3. _____ Animal |

Describe the animal's feet.
1. _____
2. _____
3. _____

How do these feet help the animal?
1. _____
2. _____
3. _____

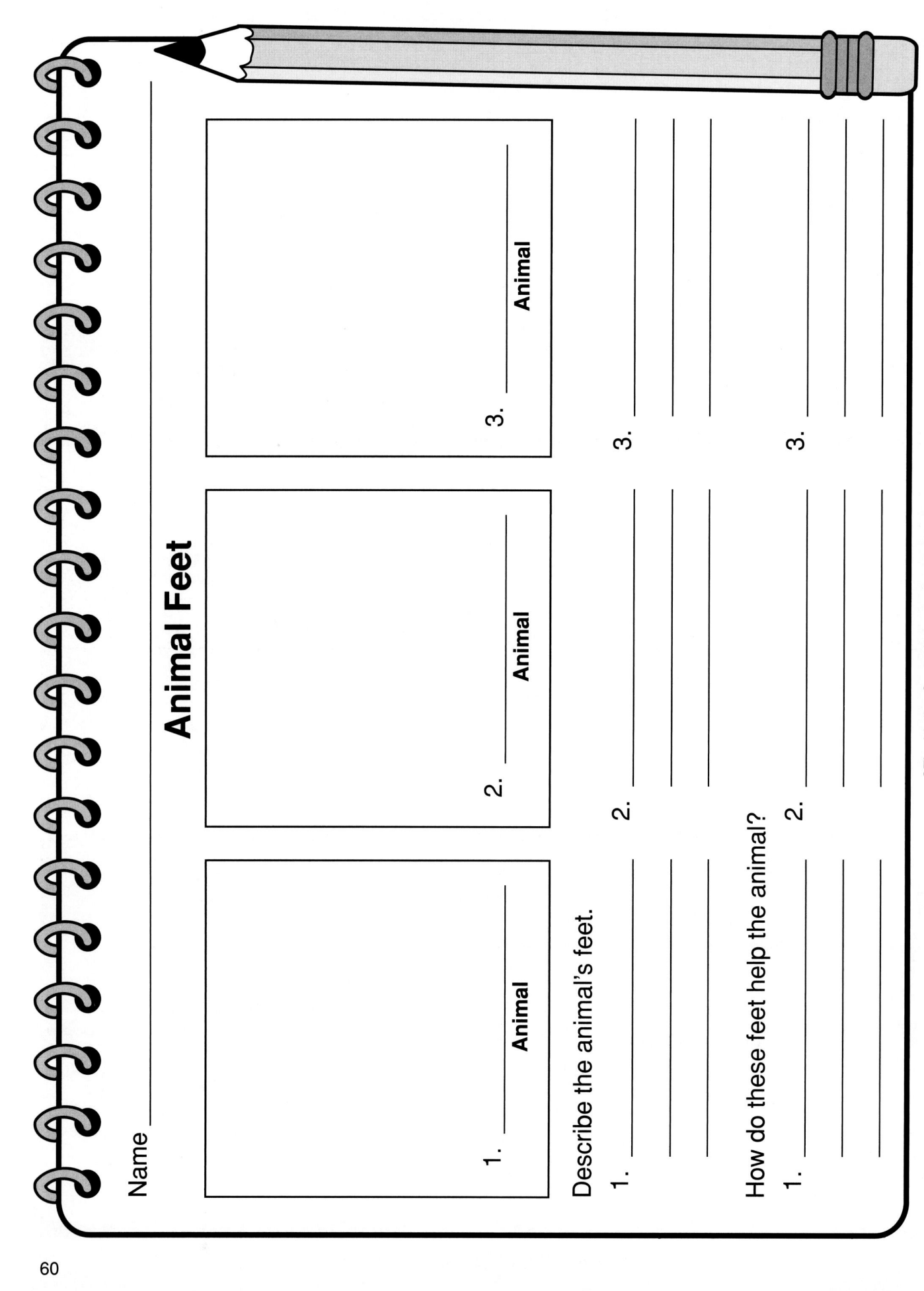

©The Mailbox® • *Science in a Box* • TEC60894

Note to the teacher: Use with the center idea on page 59.

Animal Cards
Use with the activity and center idea on page 59.

horse	squirrel
mountain goat	beaver
eagle	duck

©The Mailbox® • Science in a Box • TEC60894

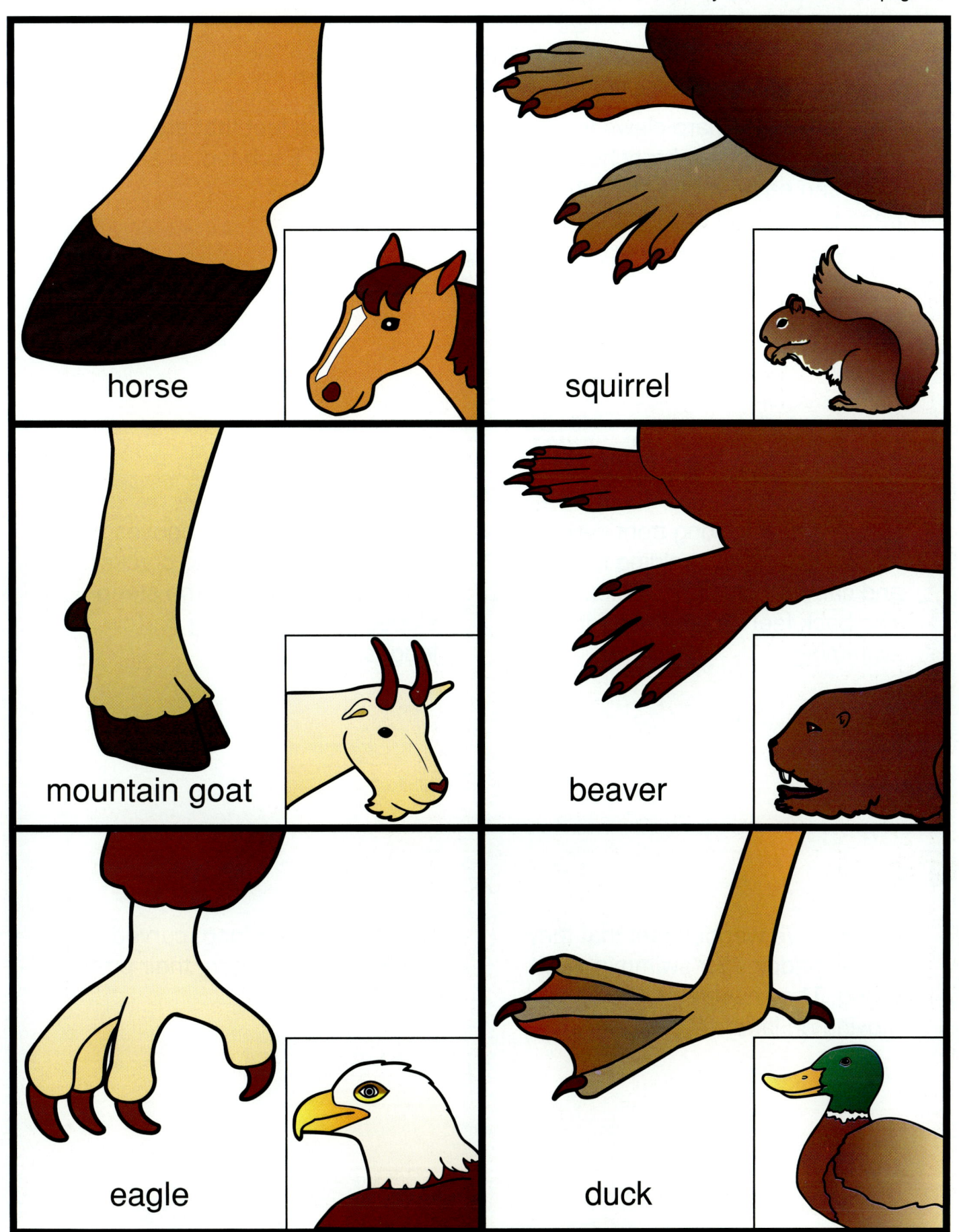

Squirrels have sharp claws to help them grab tree branches. They can easily move through trees to find food or run from enemies.	Horses have feet that are good for running. Each foot is actually a very strong toe.
Beavers have strong front claws to use for digging, holding food, and carrying mud and sticks. The back feet are webbed for swimming.	Mountain goats have hooves with sharp edges to dig into cracks in rocks. They can climb steep slopes and walk along narrow ledges.
Ducks have webbed feet that they use as paddles for swimming and diving. They have a harder time moving on land.	Eagles use their long, curved claws to kill and carry their prey.

Animal Cards

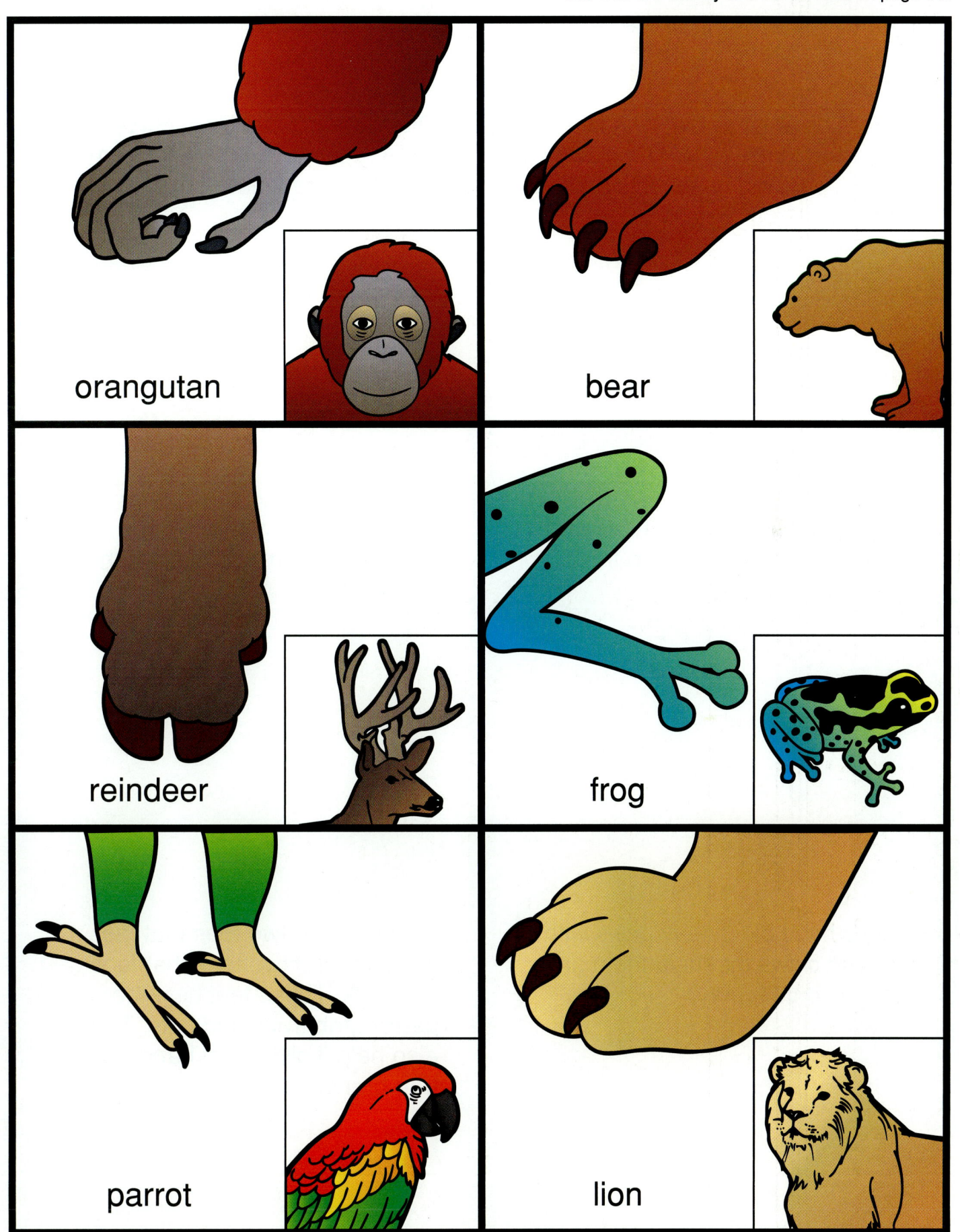

Use with the activity and center idea on page 59.

orangutan

bear

reindeer

frog

parrot

lion

Bears have large feet with five toes and long claws. The claws help a bear dig up insects and roots and tear prey.

Orangutans have long, curved fingers and toes that help them grab tree branches.

Frogs that live in trees have sticky pads on their feet to help them climb.

Reindeer have large hooves that keep them from sinking into the snow.

Lions have big, heavy paws with curved claws that help them grab and hold their prey.

Parrots have two toes that face the front and two toes that face the back. These feet help the parrot grab fruits and nuts. They also help it climb and hang from tree branches.

ANIMAL MOUTHS

Objective: to examine how an animal's mouth helps it get food

Materials: pages 65 and 67–70, supply of 6" white paper plates and 12" x 18" sheets of construction paper

Teacher preparation:
1. Follow the directions on page 3 to assemble the box.
2. Laminate this page and the animal cards on page 67 for durability. Cut apart the animal cards.
3. For each group of four students, make one copy of page 69 and the fact strips and labels on page 70.
4. Follow the directions on page 69 to make a sample of each animal mouth shown.
5. Place this page, the animal cards, the copies, and the mouth samples in the box.

Activity

Begin by having students describe characteristics of different animals' mouths. Then, ask students why they think animals' mouths are not all the same. Explain that animals' mouths have adapted to help them get the food they need to survive. Share the animal cards with students, pointing out characteristics of the different mouths. Have students predict how each animal uses its mouth to get food. Next, give each group of four students four paper plates, a copy of pages 69 and 70, and a sheet of construction paper. Have the groups use the steps on page 69 to create each animal mouth. Then have them cut out the fact strips and labels and match the cutouts to the mouths they describe. Use the answer key to check each group's work. Finally, have each group glue its project onto the construction paper.

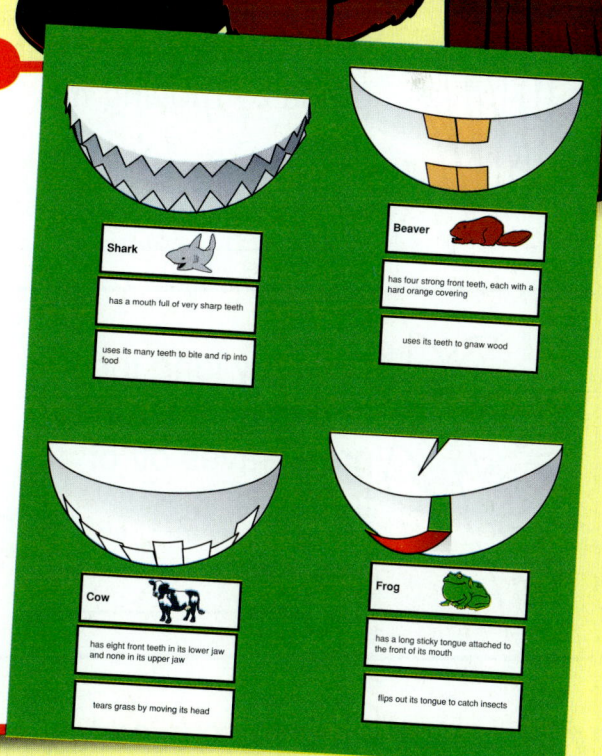

This Is Why

The characteristics of an animal's mouth determine what food an animal can eat. Animals adapt to their environment in order to get the food they need to survive.

Center Idea

Place the animal cards at a center along with a class supply of page 66. A child uses the animal cards to help him complete the page as directed. To check his work, he turns over the cards. The number of the correct answer is in the bottom left corner of the card.

©The Mailbox® • Science in a Box • TEC60894

Name _____

Animal Mouths

I use my pointed beak to remove cases from seeds.
1. _____

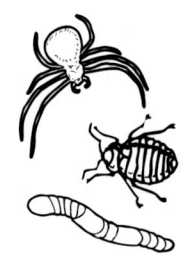

I use my sticky tongue to catch bugs.
2. _____

I only have front teeth in my lower jaw. I tear grass by moving my head.
3. _____

I use my long tube to suck nectar from flowers.
4. _____

I use my long tongue and bottom teeth to grab leaves off of trees.
5. _____

I use my four strong front teeth to gnaw wood.
6. _____

I use my curved teeth to grab small animals and pull them into my mouth.
7. _____

I use my mouth full of teeth to bite and tear into fish.
8. _____

Animal Cards

Use with the activity and center idea on page 65.

bird	beaver
snake	giraffe
cow	frog
shark	butterfly

6. A beaver has four strong front teeth. Each one has a hard orange covering. A beaver uses these teeth to gnaw wood. They will not wear out because they never stop growing. A beaver uses flat teeth in the back of its mouth for chewing.

1. A bird may have a short, sharp beak. It may use its beak to remove the hard cases from seeds before it swallows them.

5. A giraffe does not have top front teeth. It uses its long tongue and bottom teeth to grab leaves from trees. The giraffe chews with the flat teeth in the back of its mouth.

7. A snake has small, curved teeth. It uses them to grab and pull food into its mouth. A snake does not chew its food. It must swallow its food whole.

2. A frog may have a long, sticky tongue attached to the front of its mouth. It flips out its tongue to catch a bug. Then it flicks the food into its mouth and swallows it whole.

3. A cow has eight front teeth in its lower jaw. It has no top front teeth. To tear grass, a cow moves its head. A cow chews with the teeth in the back of its mouth.

4. A butterfly doesn't chew its food. It uses a long tube (called a *proboscis*) to suck the nectar from flowers.

8. A shark has many rows of teeth. It uses its teeth to bite into and tear food. New teeth take the place of older teeth often.

Shark

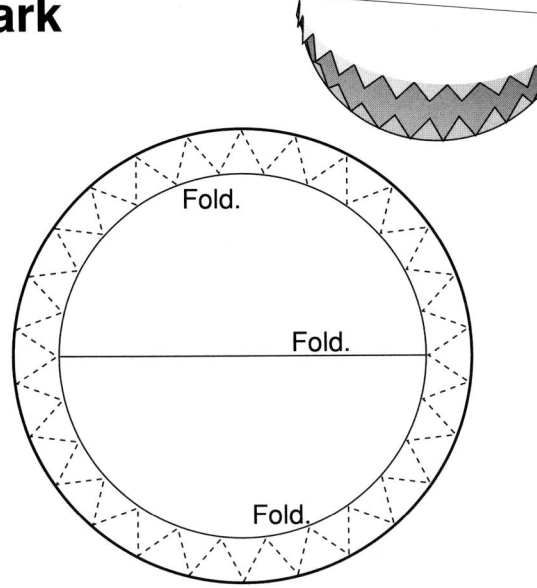

1. Cut. Use the dotted lines as a guide.
2. Fold the plate in half.
3. Fold the teeth. Use the solid lines as a guide.

Cow

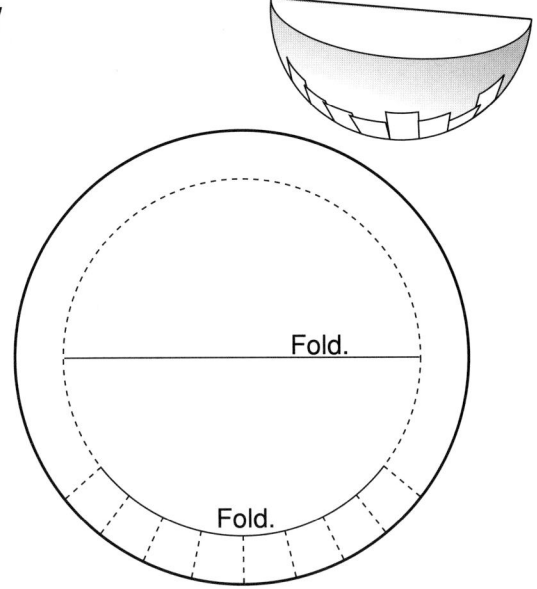

1. Cut. Use the dotted lines as a guide.
2. Fold the plate in half.
3. Fold the teeth. Use the solid lines as a guide.

Frog

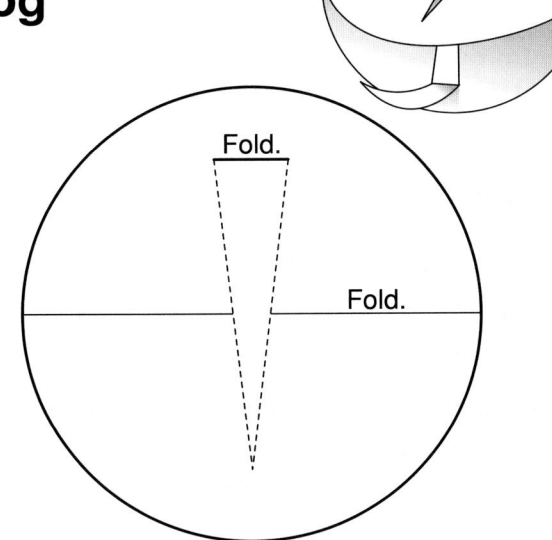

1. Cut. Use the dotted lines as a guide.
2. Color both sides of the tongue red.
3. Fold the plate in half.
4. Roll the tongue so that it curves.

Beaver

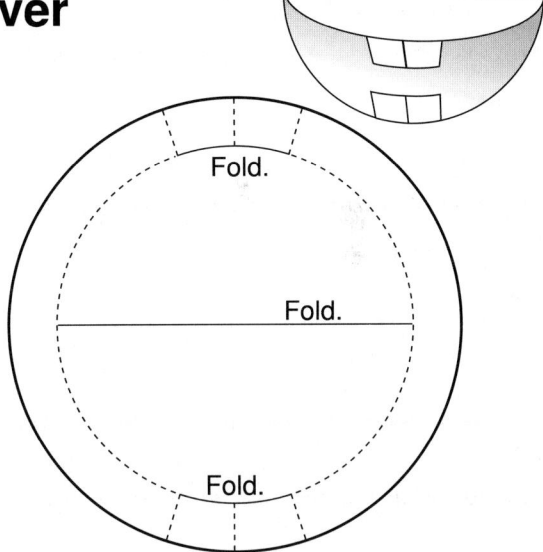

1. Cut. Use the dotted lines as a guide.
2. Color the four teeth orange.
3. Fold the plate in half.
4. Fold the teeth. Use the solid lines as a guide.

©The Mailbox® • *Science in a Box* • TEC60894

Note to the teacher: Use with the activity on page 65.

Fact Strips and Labels
Use with the activity on page 65.

has a mouth full of very sharp teeth	has four strong front teeth, each with a hard orange covering
has a long sticky tongue attached to the front of its mouth	has eight front teeth in its lower jaw and none in its upper jaw
uses its teeth to gnaw wood	flips out its tongue to catch bugs
tears grass by moving its head	uses its many teeth to bite and rip into food
Beaver	**Cow**
Frog	**Shark**

©The Mailbox® • Science in a Box • TEC60894

- -

Answer Key

Frog
— has a long sticky tongue attached to the front of its mouth
— flips out its tongue to catch bugs

Cow
— has eight front teeth in its lower jaw and none in its upper jaw
— tears grass by moving its head

Beaver
— has four strong front teeth, each with a hard orange covering
— uses its teeth to gnaw wood

Shark
— has a mouth full of very sharp teeth
— uses its many teeth to bite and rip into food

ANIMAL WINGS

Objective: to explore the importance, functions, and differences of various animal wings

Materials: pages 71–73

Teacher preparation:
1. Follow the directions on page 3 to assemble the box.
2. Laminate this page and the animal cards on page 73 for durability.
3. Cut apart the animal cards.
4. Make a class supply of page 72.
5. Place this page, the animal cards, and the copies inside the box. If desired, also store the remaining pages from this unit in the box.

Activity

Begin by asking students to look closely at the animal cards to see what the pictured animals have in common *(wings)*. Next, write the following headings on the board: "Bats," "Birds," and "Insects." Explain that these are the only groups of animals that have wings. Guide students to match each animal card to one of the groups. Next, have students discuss how the pictured wings are similar and different. Explain that bird wings are made of bones and feathers and vary in shape (such as long or short, narrow or wide, rounded or pointed). Insect wings are made of a thin skin called *membrane*. Some are transparent and the veins can be seen running through them (as in flies and grasshoppers). Other insect wings are covered in tiny scales (such as those on butterflies). Finally, point out that a bat is the only mammal with wings. Its wings are made of bones covered in two thin layers of skin. Follow up by giving each child a copy of page 72 to complete as directed.

This Is Why

An animal's wings help it fly. Only three groups of animals have wings: insects, bats, and birds. Each group's wings are constructed differently. The wings of birds have different shapes, allowing some to fly at different speeds and make sharp movements while in flight.

Center Idea

Place scissors and glue at a center along with a class supply of pages 75 and 76. A student cuts out the cards on page 75 and matches each bird to its shape description on page 76. He completes the chart by gluing each bird card in place.

©The Mailbox® • Science in a Box • TEC60894

71

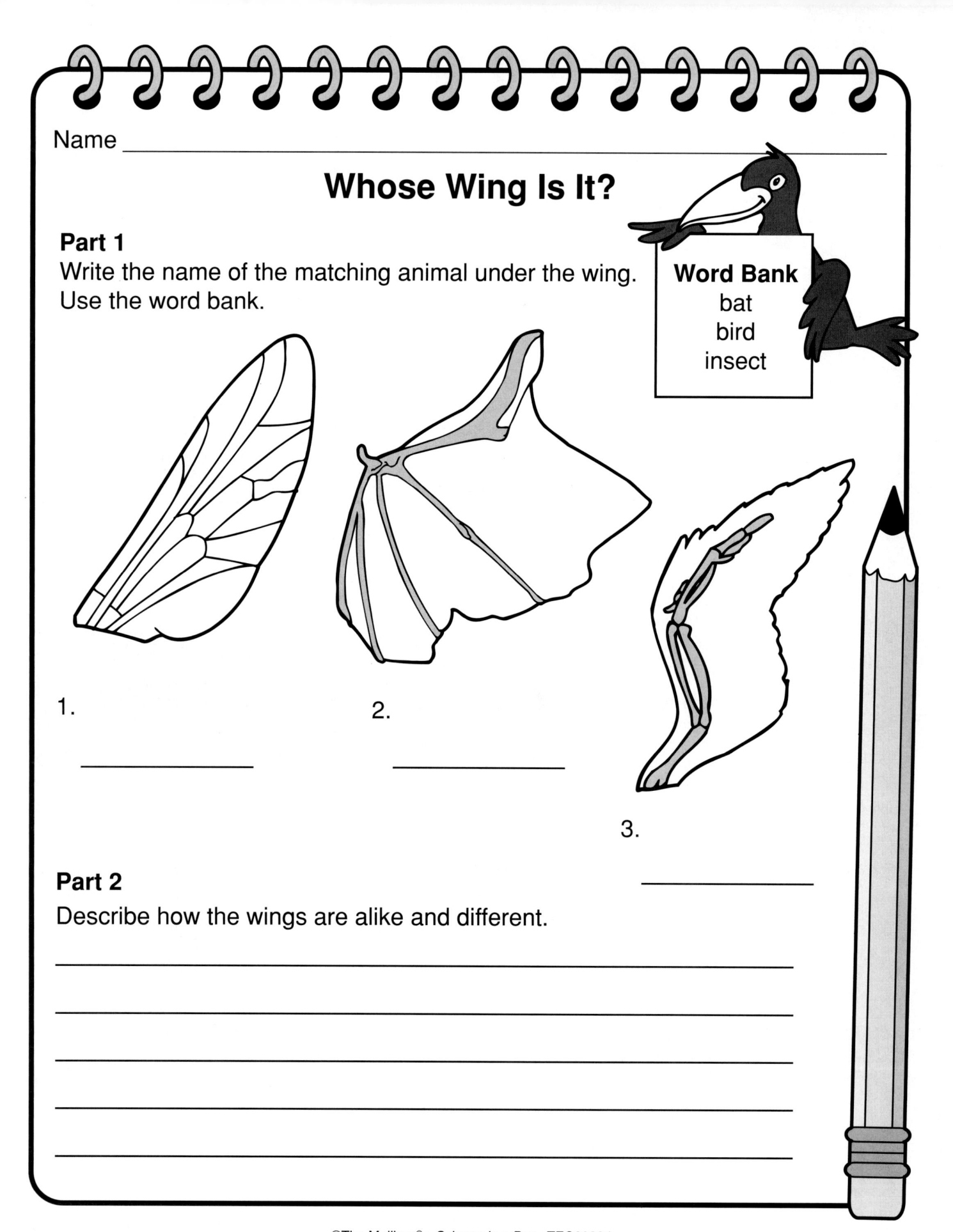

Name _____

Whose Wing Is It?

Part 1
Write the name of the matching animal under the wing. Use the word bank.

Word Bank
bat
bird
insect

1. _____

2. _____

3. _____

Part 2
Describe how the wings are alike and different.

_____
_____
_____
_____
_____

©The Mailbox® • Science in a Box • TEC60894

Note to the teacher: Use with the activity on page 71.

Animal Cards
Use with the activity on page 71.

eagle	grasshopper	bat
blue jay	robin	dragonfly
fly	butterfly	chickadee

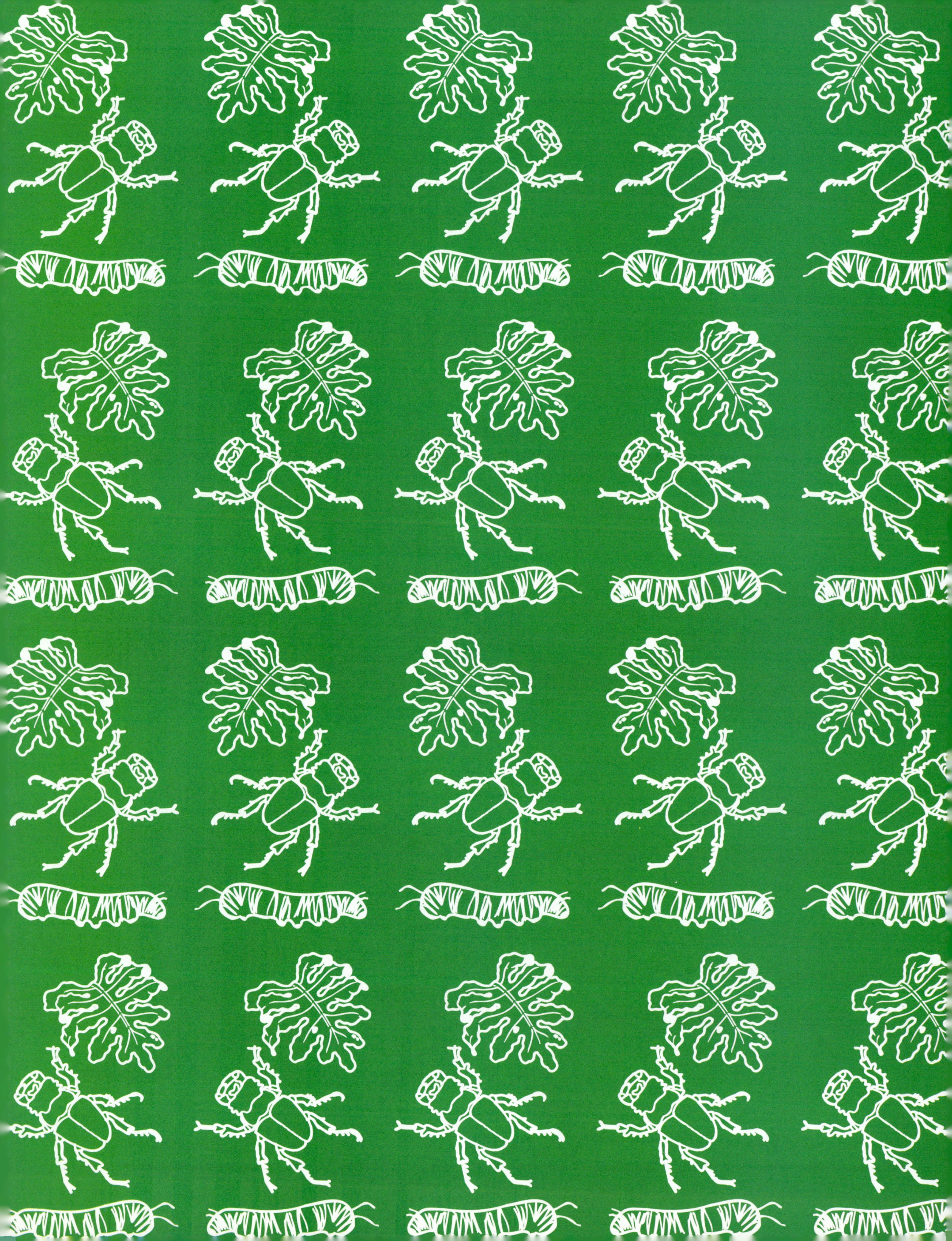

Bird Cards
Use with the center idea on page 71.

swift	pheasant	stork
albatross	hawk	frigate bird
swallow	owl	

©The Mailbox® • *Science in a Box* • TEC60894

Answer Key

Page 72
Part 1
1. insect
2. bat
3. bird

Part 2
Descriptions may vary.

Page 76
Order of cards in each category may vary.
Long and Pointed: albatross, frigate bird
Long and Wide: hawk, stork
Short and Rounded: owl, pheasant
Small and Narrow: swift, swallow

Wing Shapes Name _____

Long and Pointed Many birds with these wings glide in the wind over the ocean.		
Long and Wide Birds with these wings soar on rising air currents.		
Short and Rounded Birds with these wings take off quickly and may make sharp turns around trees.		
Small and Narrow Birds with these wings can fly very fast.		

©The Mailbox® • Science in a Box • TEC60894

Note to the teacher: Use with the center idea on page 71.

ANIMAL HABITATS

Objective: to match animals to their habitats

Materials: pages 77–79

Teacher preparation:
1. Follow the directions on page 3 to assemble the box.
2. Laminate this page and the animal cards on page 79 for durability. Cut apart the cards.
3. Make a class supply of page 78.
4. Place this page, the animal cards, and the copies inside the box. If desired, also store the remaining page from this unit in the box.

Activity

Begin by asking each student to think about her favorite wild animal. Have a few student volunteers share their choices. Next, ask those same students to identify where each of the named animals lives. Explain that the area where an animal lives is called its *habitat*. Write five habitats on the board as shown. Distribute one animal card to each student or pair of students. Have each student think about where his animal lives and then tape the card under the matching habitat. Next, ask students to think about what would happen if an animal wanted to change habitats. For example, they might consider what would happen if a turkey attempted to live in the desert. Point out that some animals have very different needs. A habitat that is good for one animal may not be good for another. A turkey might not survive in the dry desert. Discuss a few more examples and then have each student complete a copy of page 78.

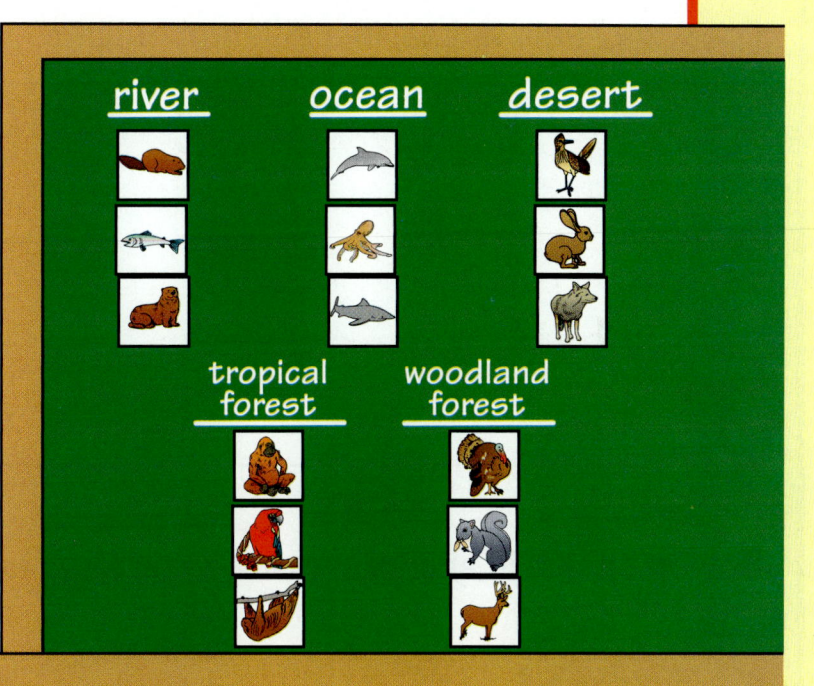

This Is Why

The place where an animal lives is its habitat. Animals live where they can find what they need to survive.

Center Idea

Laminate the habitat cards on page 81 and place them in a center along with the animal cards from page 79. Invite pairs of students to visit the center to play a Memory-style game, matching each animal to its habitat.

©The Mailbox® • Science in a Box • TEC60894

Name _____

Animal Habitats

Draw an animal in its habitat.

Draw the same animal in a different habitat.

Do you think this animal can survive in this new habitat? Why or why not?

_____
_____
_____
_____

©The Mailbox® • *Science in a Box* • TEC60894

Note to the teacher: Use with the activity on page 77.

Animal Cards

Use with the activity and center idea on page 77.

beaver	trout	otter
dolphin	octopus	shark
roadrunner	jackrabbit	coyote
orangutan	macaw	sloth
turkey	squirrel	deer

©The Mailbox • Science in a Box • TEC60894

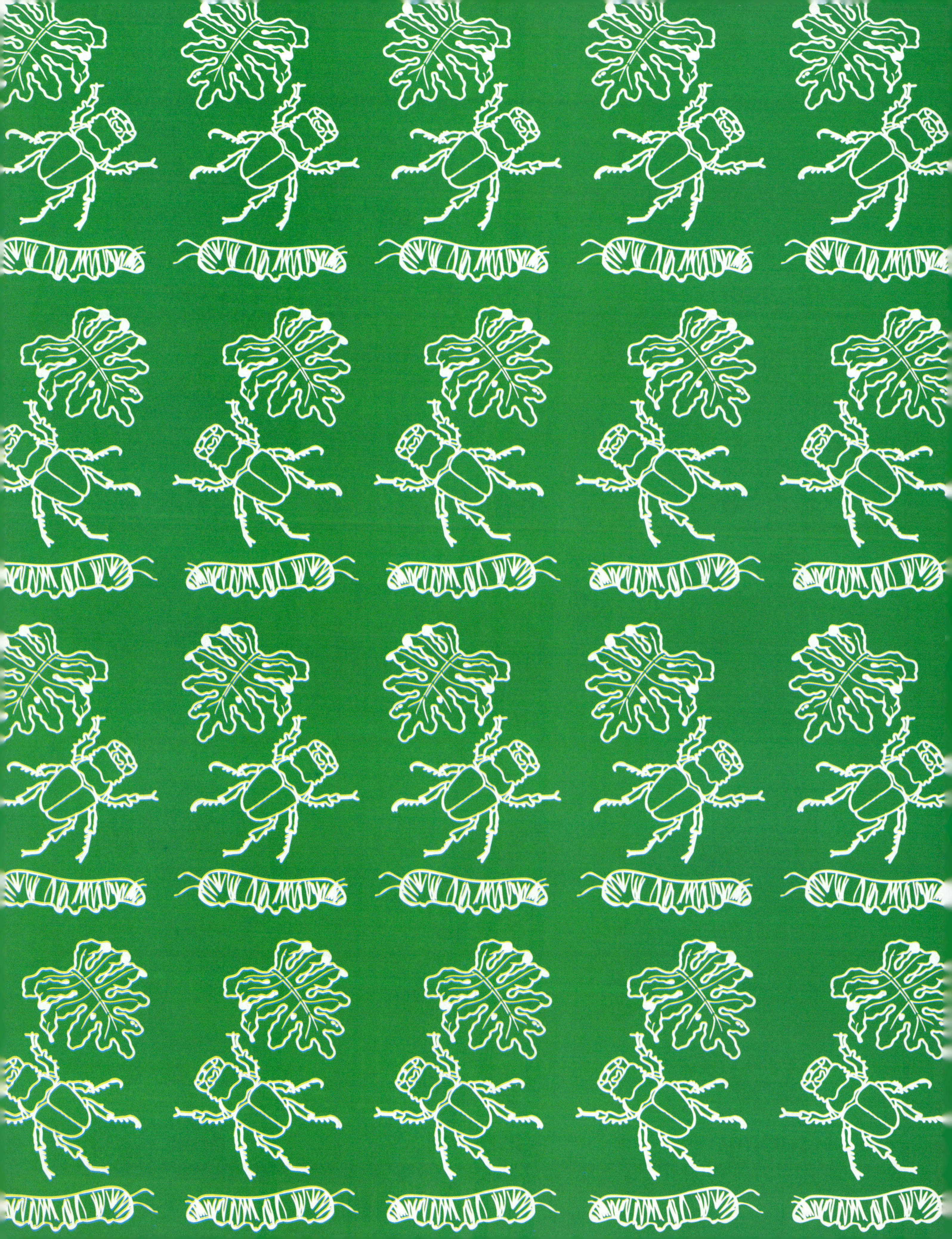

Habitat Cards
Use with the center idea on page 77.

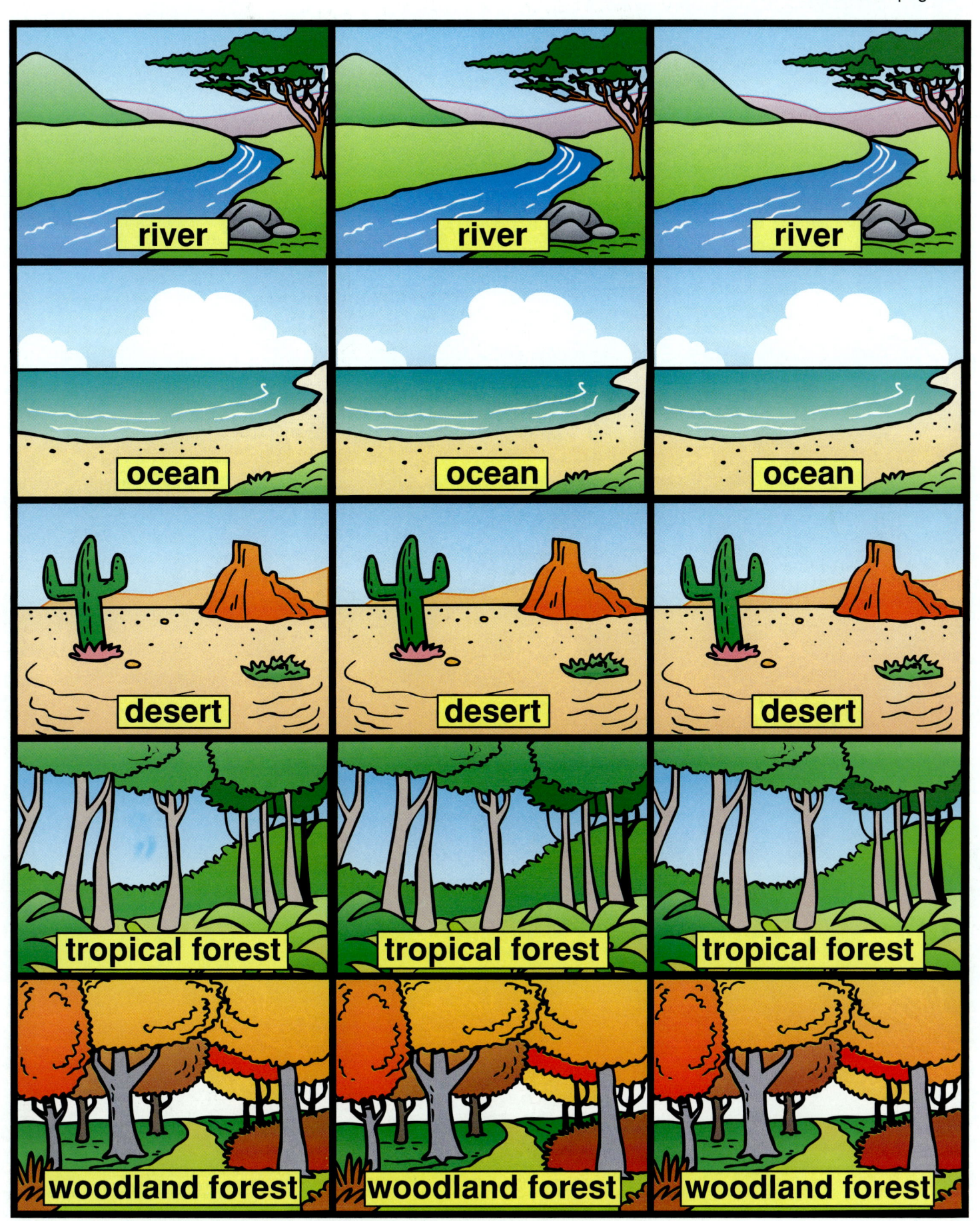

river	river	river
ocean	ocean	ocean
desert	desert	desert
tropical forest	tropical forest	tropical forest
woodland forest	woodland forest	woodland forest

©The Mailbox® • *Science in a Box* • TEC60894

ANIMAL ADAPTATIONS

Objective: to understand how polar bears have adapted to their environment

Materials: pages 83–86

Teacher preparation:
1. Follow the directions on page 3 to assemble the box.
2. Laminate this page for durability.
3. Make a copy of pages 85 and 86. Cut apart the cards.
4. Make a class supply of page 84.
5. Place this page, the cards, and the copies inside the box. If desired, also store the remaining pages from this unit in the box.

Activity

Begin by brainstorming a class list of things that students know about polar bears. Next, ask students what it must be like living in the icy Arctic where polar bears live. Discuss what aspects of that environment might make it difficult to survive. Then tape the three illustrated cards from page 85 to the board as shown. Explain that polar bears have adapted over time to living in the Arctic. Read each of the remaining cards and have students decide whether that characteristic helps a polar bear hunt for food, stay warm, or both. Tape each card under the matching category. Afterward, give each child a copy of page 84 to complete as a follow-up.

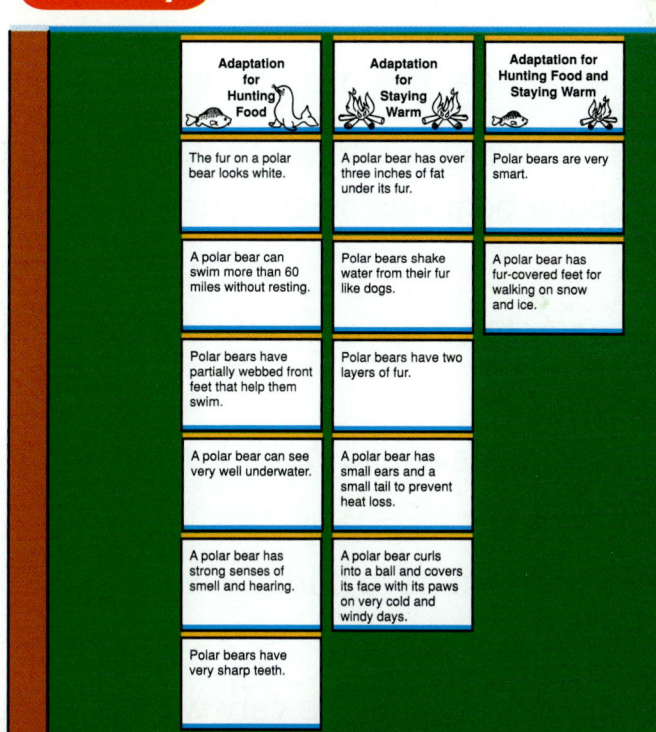

Adaptation for Hunting Food	Adaptation for Staying Warm	Adaptation for Hunting Food and Staying Warm
The fur on a polar bear looks white.	A polar bear has over three inches of fat under its fur.	Polar bears are very smart.
A polar bear can swim more than 60 miles without resting.	Polar bears shake water from their fur like dogs.	A polar bear has fur-covered feet for walking on snow and ice.
Polar bears have partially webbed front feet that help them swim.	Polar bears have two layers of fur.	
A polar bear can see very well underwater.	A polar bear has small ears and a small tail to prevent heat loss.	
A polar bear has strong senses of smell and hearing.	A polar bear curls into a ball and covers its face with its paws on very cold and windy days.	
Polar bears have very sharp teeth.		

This Is Why

Animals that live in polar regions have developed ways of life and body characteristics that allow them to survive in the extreme cold weather. Polar bears are well adapted to life in the Arctic.

Center Idea

Make a class supply of page 87 and one copy of the center directions on page 88. Place them in a center along with construction paper, scissors, and glue. A child follows the directions to complete a polar bear puzzle.

©The Mailbox® • Science in a Box • TEC60894

Name _____

Amazing Bears

Read the fact.
Color the ice by the code.

Color Code
helps the bear hunt food = blue
helps the bear stay warm = red

1. Polar bear fur looks white. It blends in with the snow and ice.

2. Polar bears can smell food as far as ten miles away.

3. A polar bear has over three inches of fat under its fur.

4. A polar bear has small ears and a small tail.

5. A polar bear can swim more than 60 miles without resting.

6. Polar bears shake water from their fur like dogs.

7. Polar bears have partially webbed front feet that help them swim.

8. A polar bear can see very well underwater.

9. Polar bears have two layers of fur.

10. Polar bears have very sharp teeth.

©The Mailbox® • Science in a Box • TEC60894

Note to the teacher: Use with the activity on page 83.

Polar Bear Cards
Use with the activity on page 83.

Adaptation for Hunting Food	Adaptation for Staying Warm
Adaptation for Hunting Food and Staying Warm	Polar bears are very smart.
The fur on a polar bear looks white.	A polar bear has over three inches of fat under its fur.
A polar bear has small ears and a small tail to prevent heat loss.	A polar bear can swim more than 60 miles without resting.

Polar Bear Cards

Use with the activity on page 83.

Polar bears shake water from their fur like dogs.	A polar bear has fur-covered feet for walking on snow and ice.
Polar bears have partially webbed front feet that help them swim.	A polar bear can see very well underwater.
Polar bears have two layers of fur.	A polar bear curls into a ball and covers its face with its paws on very cold and windy days.
A polar bear has strong senses of smell and hearing.	Polar bears have very sharp teeth.

Polar Bear Puzzle Pieces
Use with the center idea on page 83.

1. Polar bears are well _____ to living in the _____.

2. They have two layers of _____ and a thick layer of _____ to keep warm.

3. Polar bears have very sharp _____ for hunting _____.

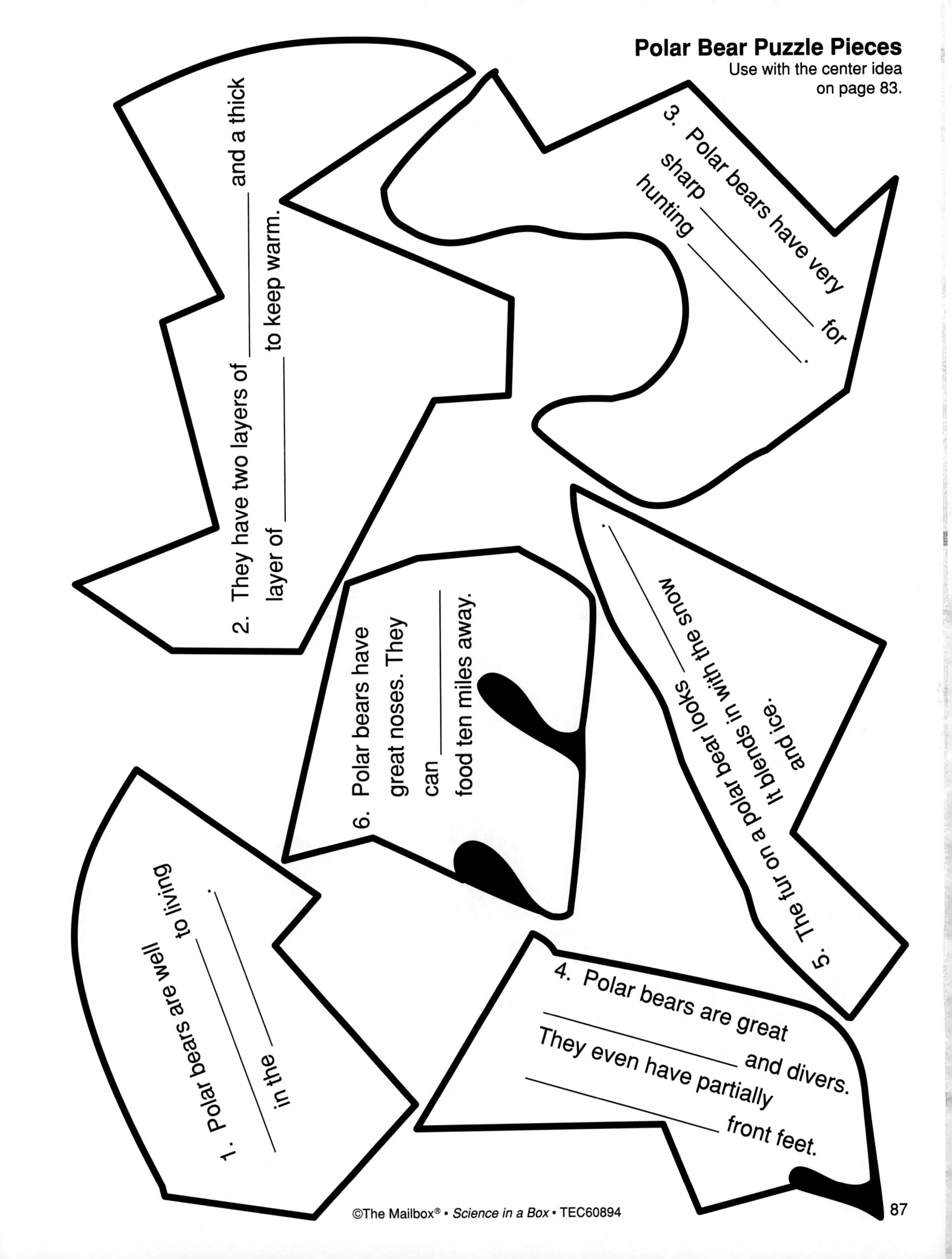

4. Polar bears are great _____ and divers. They even have partially _____ front feet.

5. The fur on a polar bear looks _____. It blends in with the snow and ice.

6. Polar bears have great noses. They can _____ food ten miles away.

©The Mailbox® • Science in a Box • TEC60894

Center Directions
Use with the center idea on page 83.

Polar Bear Puzzle

Directions

1. Complete each sentence on a puzzle piece. Use the word bank.

2. Cut apart the puzzle pieces.

3. Put them together to form a polar bear.

4. Glue your polar bear puzzle to a sheet of construction paper.

Word Bank

fur	fat
teeth	Arctic
adapted	white
smell	prey
webbed	swimmers

©The Mailbox® • Science in a Box • TEC60894

Answer Key

Page 84
1. blue
2. blue
3. red
4. red
5. blue
6. red
7. blue
8. blue
9. red
10. blue

Polar Bear Puzzle

1. Polar bears are well <u>adapted</u> to living in the <u>Arctic</u>.
2. They have two layers of <u>fur</u> and a thick layer of <u>fat</u> to keep warm.
3. Polar bears have very sharp <u>teeth</u> for hunting <u>prey</u>.
4. Polar bears are great <u>swimmers</u> and divers. They even have partially <u>webbed</u> front feet.
5. The fur on a polar bear looks <u>white</u>. It blends in with the snow and ice.
6. Polar bears have great noses. They can <u>smell</u> food ten miles away.

ANIMAL LIFE CYCLES

Objective: to compare the life cycles of a grasshopper and a frog

Materials: page 89, bottom half of page 91, page 93, 6" x 12" strip of construction paper for each child

Teacher preparation:
1. Follow the directions on page 3 to assemble the box.
2. Laminate this page for durability.
3. Make a class supply of the black-and-white life cycle cards on page 91 and the booklet pages on page 93.
4. Place this page and the copies inside the box. If desired, also store the remaining pages from this unit in the box.

Activity

Begin by asking students how a grasshopper and a frog are alike and how they are different. After several responses, point out that another way they are different is in their life cycles. Remind students that a life cycle describes the changes that a living thing goes through during its life. To construct a life cycle booklet, give each student a 6" x 12" construction paper strip, a copy of the black-and-white life cycle cards on page 91, and a copy of the booklet pages on page 93. Have him cut apart the booklet pages along the bold lines and then cut each page along the dotted line. Next, have him stack the four pages sequentially and staple them between the folded construction paper strip. When all the booklets are constructed, instruct each student to cut out the life cycle cards. Then, as a class, use the word bank to complete each sentence and glue on the matching picture. Finally, have each student personalize and decorate his booklet cover.

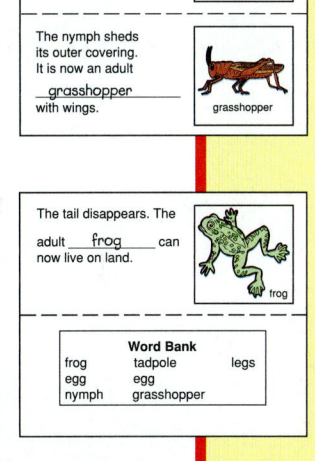

This Is Why

All living things have a life cycle. During their life cycles, both a frog and grasshopper go through *metamorphosis*, or transformation. However, their life cycles are very different.

Center Idea

Laminate the color life cycle cards on page 91. Cut out the cards and place them in a center along with the center mat from page 90 and a class supply of page 94. A student places the life cycle cards in order on the mat for each animal. (If placed in the correct order, the back of the cards will spell FROG and HOP.) Then he uses the cycles to complete page 94.

©The Mailbox® • Science in a Box • TEC60894

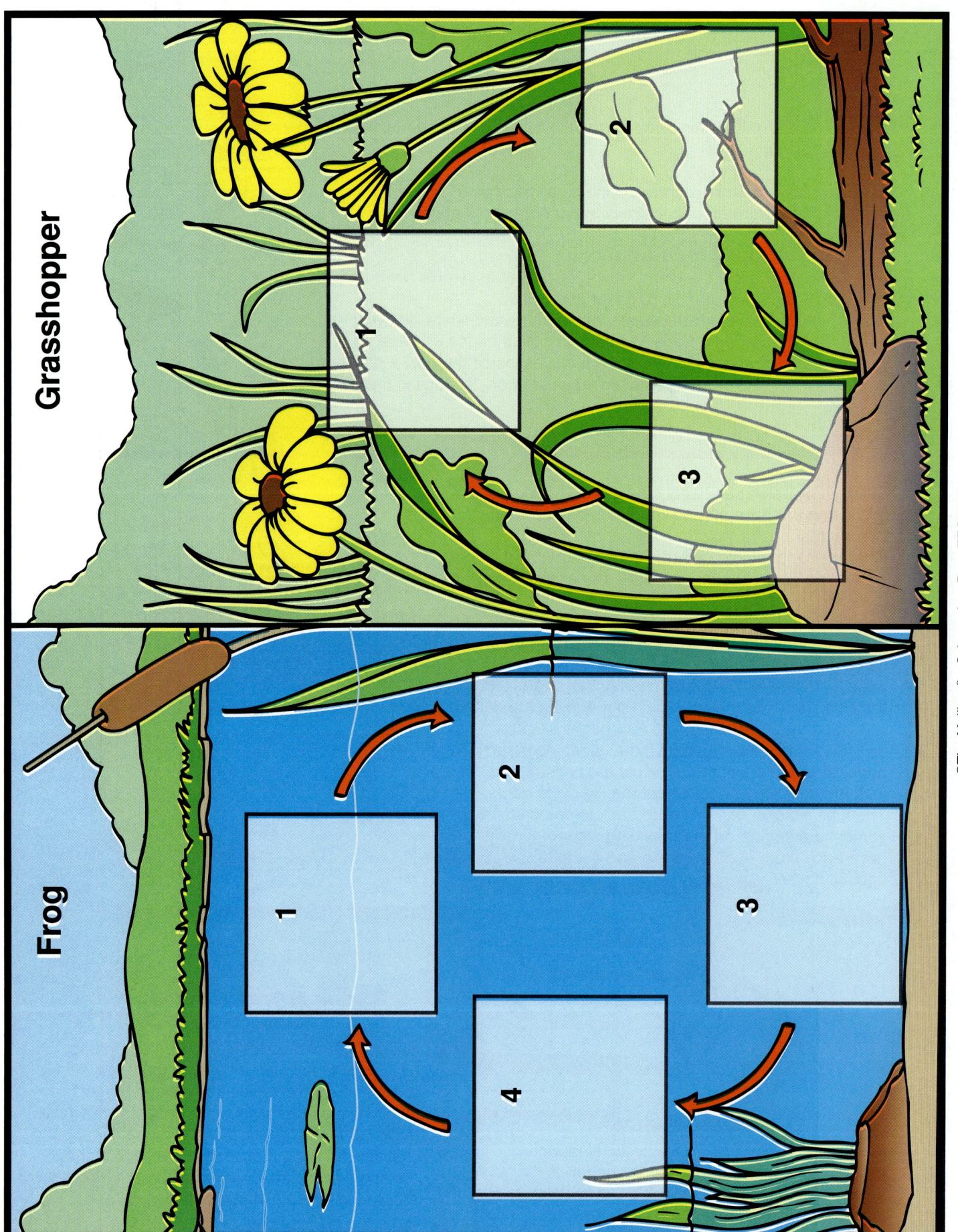

Grasshopper

Frog

Color Life Cycle Cards
Use with the center idea on page 89.

Black-and-White Life Cycle Cards
Use with the activity on page 89.

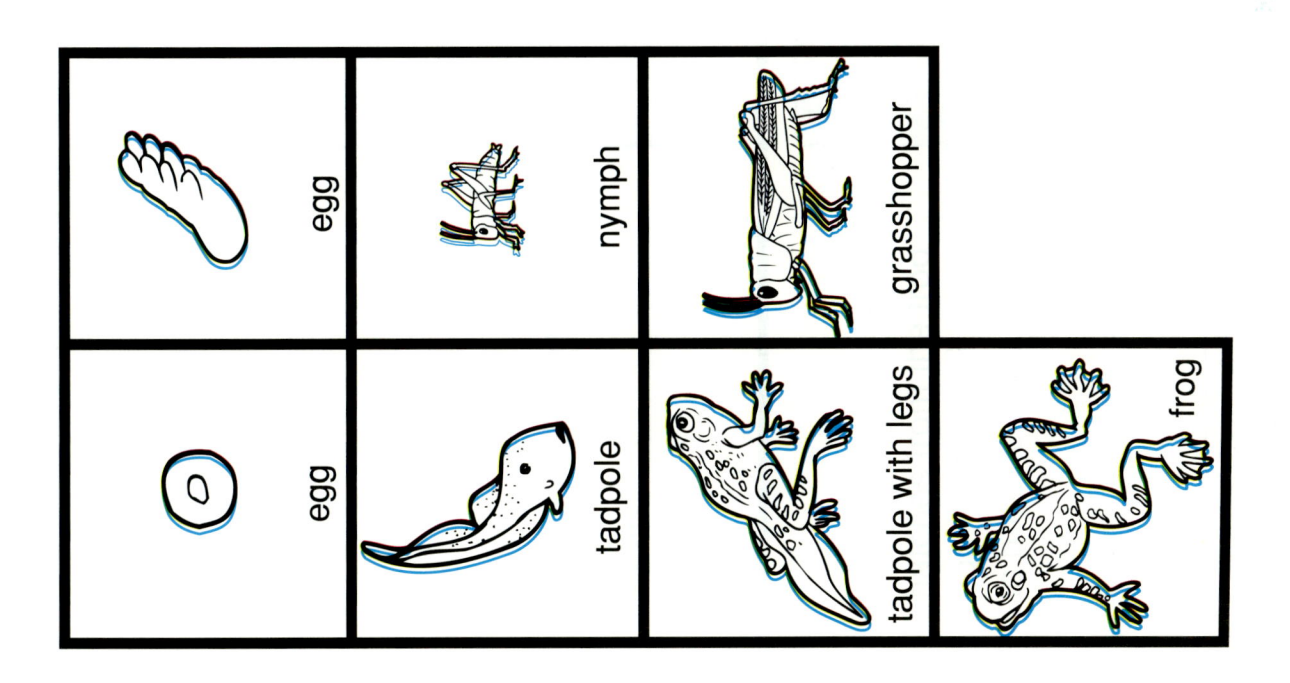

Name _____

A Chain Reaction

Draw the food chain.
Use the animals listed in the word bank.

Word Bank

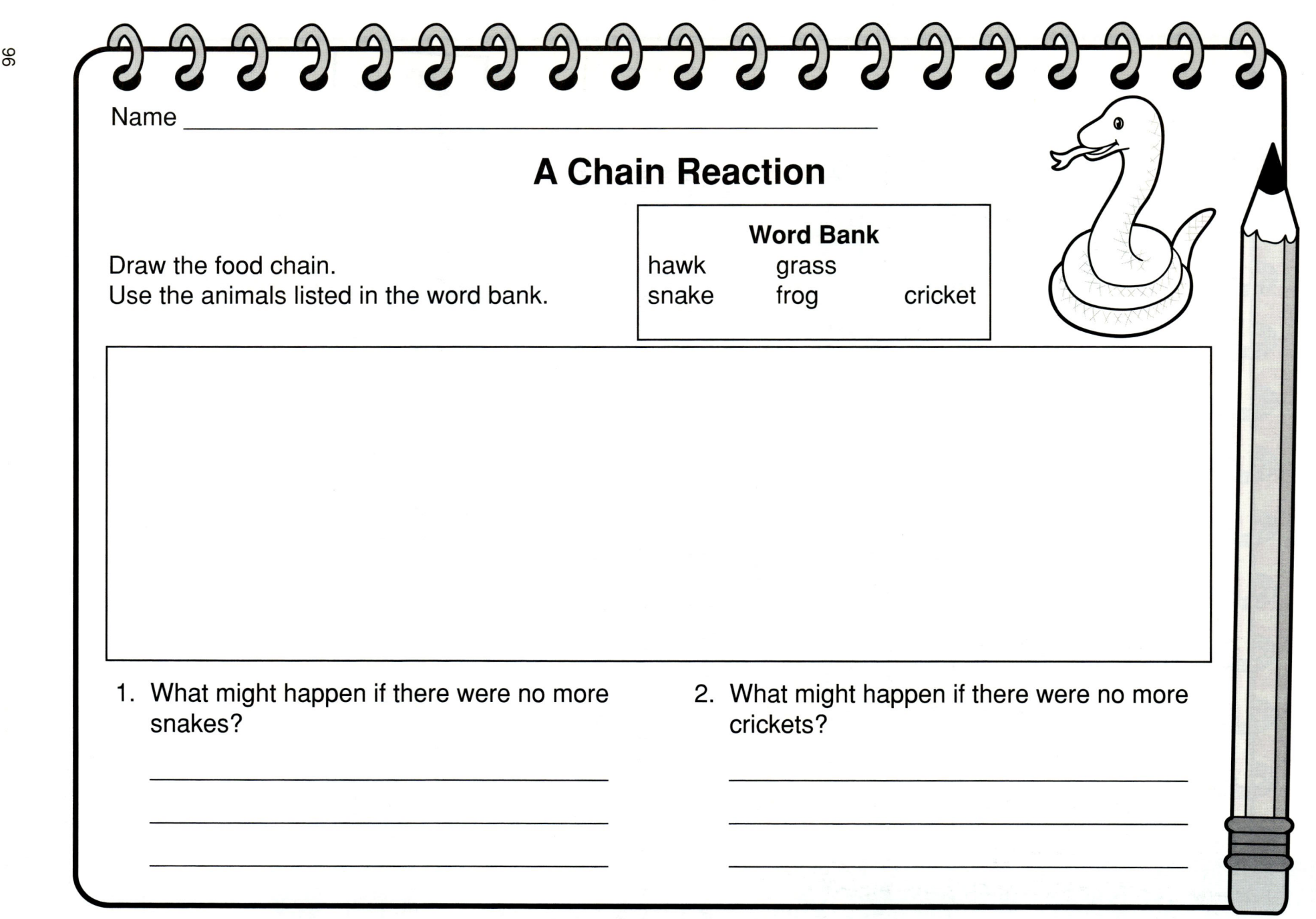

hawk   grass
snake   frog   cricket

1. What might happen if there were no more snakes?
   _____
   _____

2. What might happen if there were no more crickets?
   _____
   _____

96

©The Mailbox® • *Science in a Box* • TEC60894

Note to the teacher: Use with the activity on page 95.

Interdependence Cards
Use with the activity on page 95.

grass

cricket

frog

snake

hawk

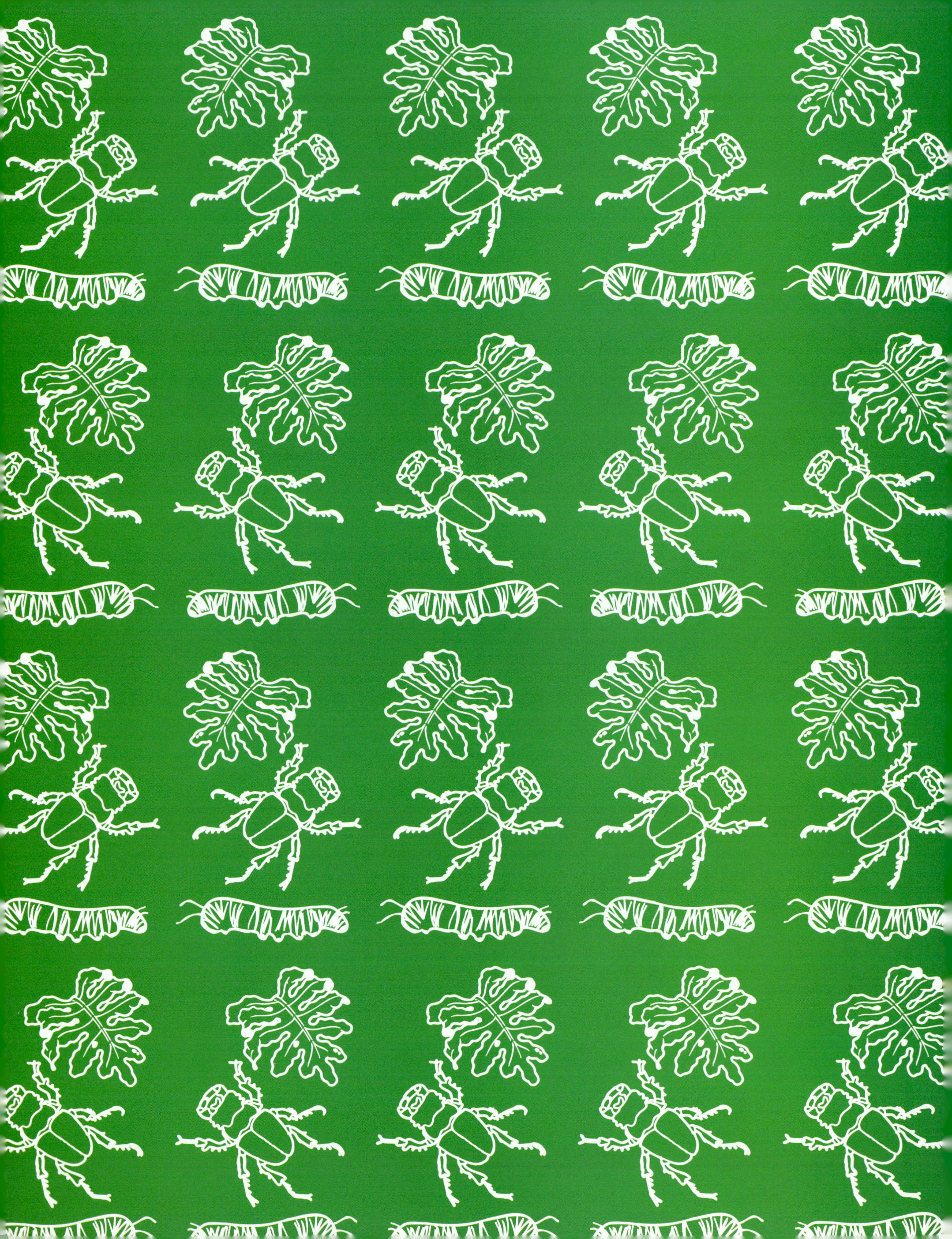

Booklet Pages
Use with the center idea on page 95.

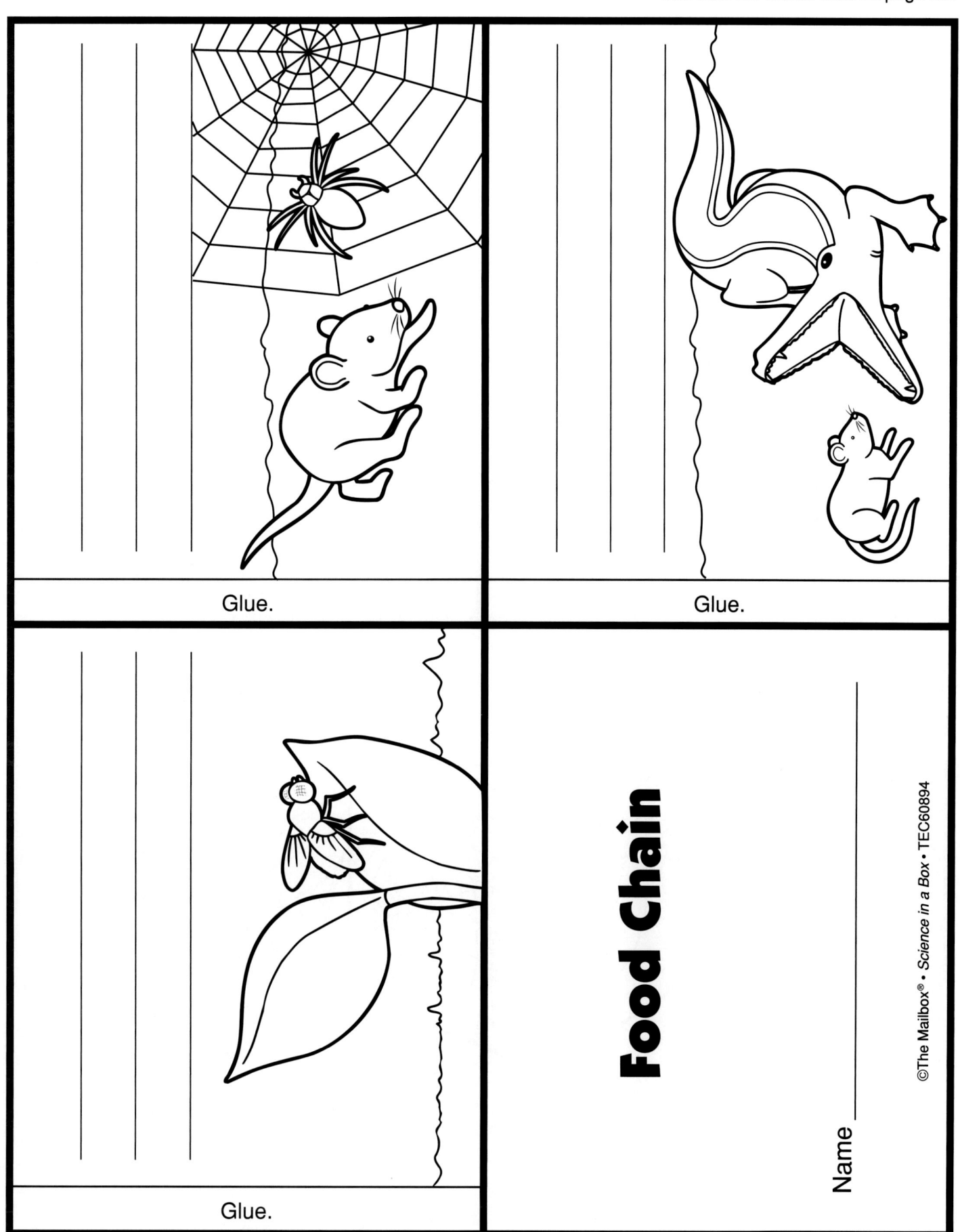

Glue.

Glue.

Glue.

Food Chain

Name

©The Mailbox® • Science in a Box • TEC60894

Booklet Pages

Use with the center idea on page 95.

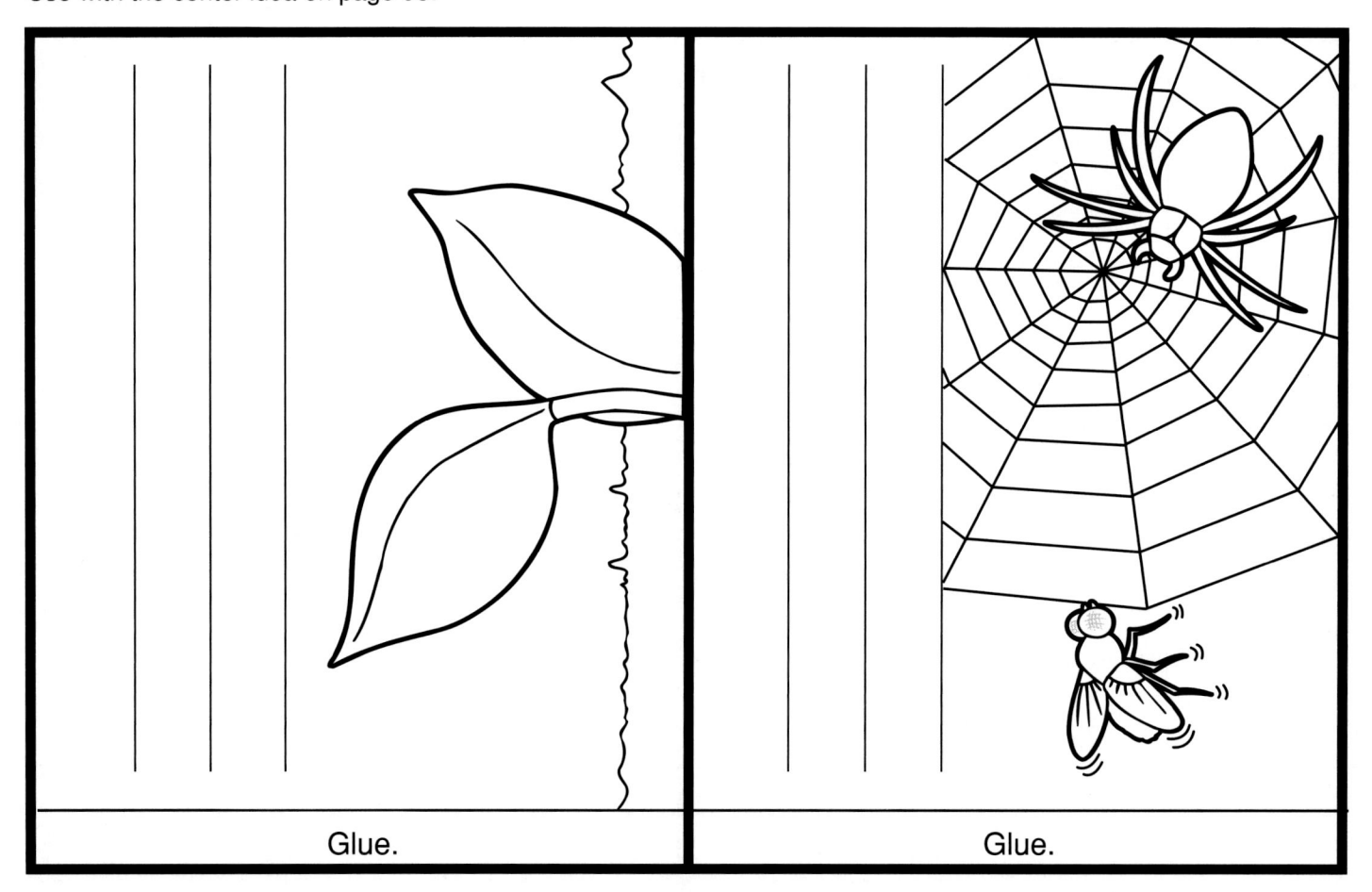

Glue.

Glue.

Finished Sample

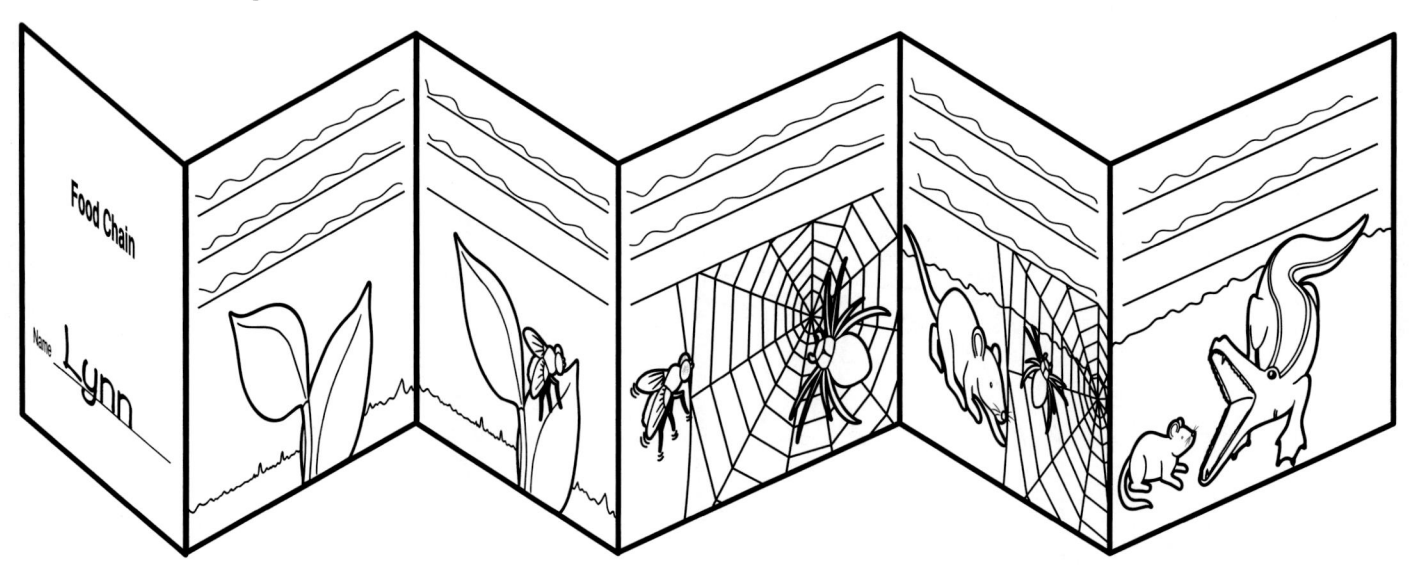

OBSERVING WEATHER

Objective: to observe and record changes in the weather

Materials: pages 101–102

Teacher preparation:
1. Follow the directions on page 3 to assemble the box.
2. Laminate this page for durability.
3. Make a class supply of the chart on page 102.
4. Place this page and the copies in the box. If desired, also store the remaining pages from this unit in the box.

Activity

Begin by asking students to describe today's weather. Encourage them to use words that describe the conditions, such as *cold, rainy,* and *sunny.* If there is a thermometer available, have one student identify the outside temperature. Continue the discussion by asking students to describe how the weather changes during the day, during the week, and during the year.

Next, give each child a copy of the chart on page 102. Have students work in groups of four to make one booklet for the group. To make the booklet, each group member takes a turn completing the chart for one week, recording information about the weather for each day of his week. At the end of the week, he writes a few sentences telling how the weather changed during the week. After four weeks, staple each group's pages together to form a booklet. Then have the groups compare and discuss their findings.

This Is Why

Each place in the world experiences many different types of weather. Some of the different types are sunny, rainy, snowy, and windy. The weather affects the way people dress, the things they do, and the foods they eat.

Center Idea

Laminate page 103. Also make a copy of page 105 and the graph on page 106 for each pair of students. Place the copies in a center along with a pencil and a paper clip to use as the spinner. A pair of students follows the directions to fill in the calendar and then records its findings on a graph.

Weather Charts
Use with the activity on page 101.

Week 1 2 3 4	Sunday	Monday	Tuesday
Wednesday	Thursday	Friday	Saturday

_____
_____
_____

©The Mailbox® • *Science in a Box* • TEC60894

Week 1 2 3 4	Sunday	Monday	Tuesday
Wednesday	Thursday	Friday	Saturday

_____
_____
_____

©The Mailbox® • *Science in a Box* • TEC60894

What a Month!

Directions:
1. Spin the spinner.
2. Record the symbol on any day of your calendar.
3. Repeat until all the days are filled.
4. Make a graph of your results.

- Windy
- Sunny
- Cloudy
- Rainy
- Snowy

©The Mailbox® • Science in a Box • TEC60894

What a Month!

Sunday	Monday	Tuesday	Wednesday	Thursday	Friday	Saturday
				1	2	3
4	5	6	7	8	9	10
11	12	13	14	15	16	17
18	19	20	21	22	23	24
25	26	27	28	29	30	

Note to the teacher: Use with the center idea on page 101.

Graph Patterns
Use with the center idea on page 101.

A Wonderful Weather Graph

| Windy |
| Sunny |
| Snowy |
| Rainy |
| Cloudy |

©The Mailbox® • *Science in a Box* • TEC60894

A Wonderful Weather Graph

| Windy |
| Sunny |
| Snowy |
| Rainy |
| Cloudy |

©The Mailbox® • *Science in a Box* • TEC60894

SEASONS

Objective: to learn about how seasons change in different geographic locations

Materials: pages 107–109, globe, class supply of brass fasteners

Teacher Preparation:
1. Follow the directions on page 3 to assemble the box.
2. Laminate this page for durability.
3. Make a class supply of pages 108 and 109.
4. Place this page and the copies in the box. If desired, also store the remaining pages from this unit in the box.

Activity

Begin the activity by showing students a globe. As you gently spin the globe, explain to students that the earth revolves around the sun. Further explain that, as it revolves, it rotates on its axis. Tell students that as the earth moves around the sun, parts of it receive more sunlight than others. The amount of light and heat each area gets determines what season it is in that area. Next, give each student a copy of pages 108 and 109 and a brass fastener. Have the child color and cut out the model pieces. Then show the child how to assemble his model as shown.

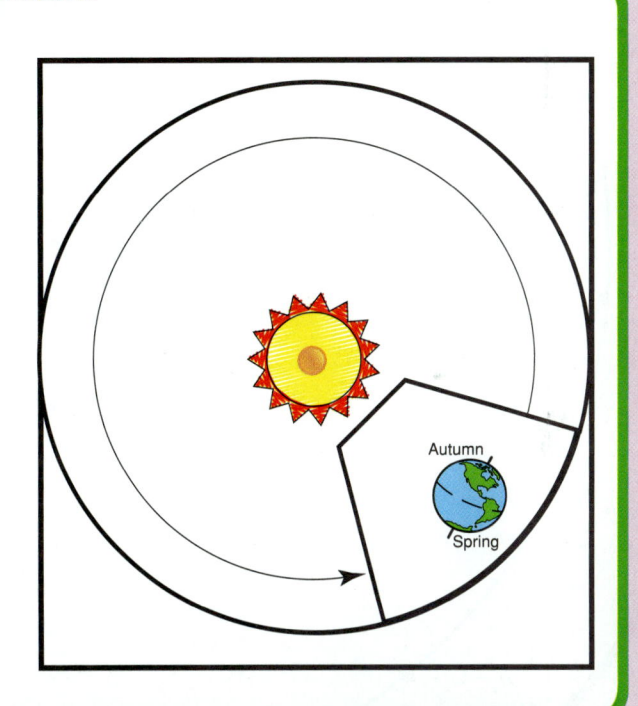

This Is Why

The earth rotates as it revolves around the sun. As the earth rotates, different places receive different amounts of light and heat. The amount of light and heat a place gets determines the season in that place.

Center Idea

Place a class supply of pages 110, 111, and 112 in a center along with scissors, crayons, and glue. A student cuts apart the booklet pages on page 110. Next, he colors and cuts apart the booklet pieces on page 111 and glues each one onto the corresponding booklet page. Then he colors and cuts apart the booklet cover on page 112. Finally, he completes the last booklet page and staples his pages in order.

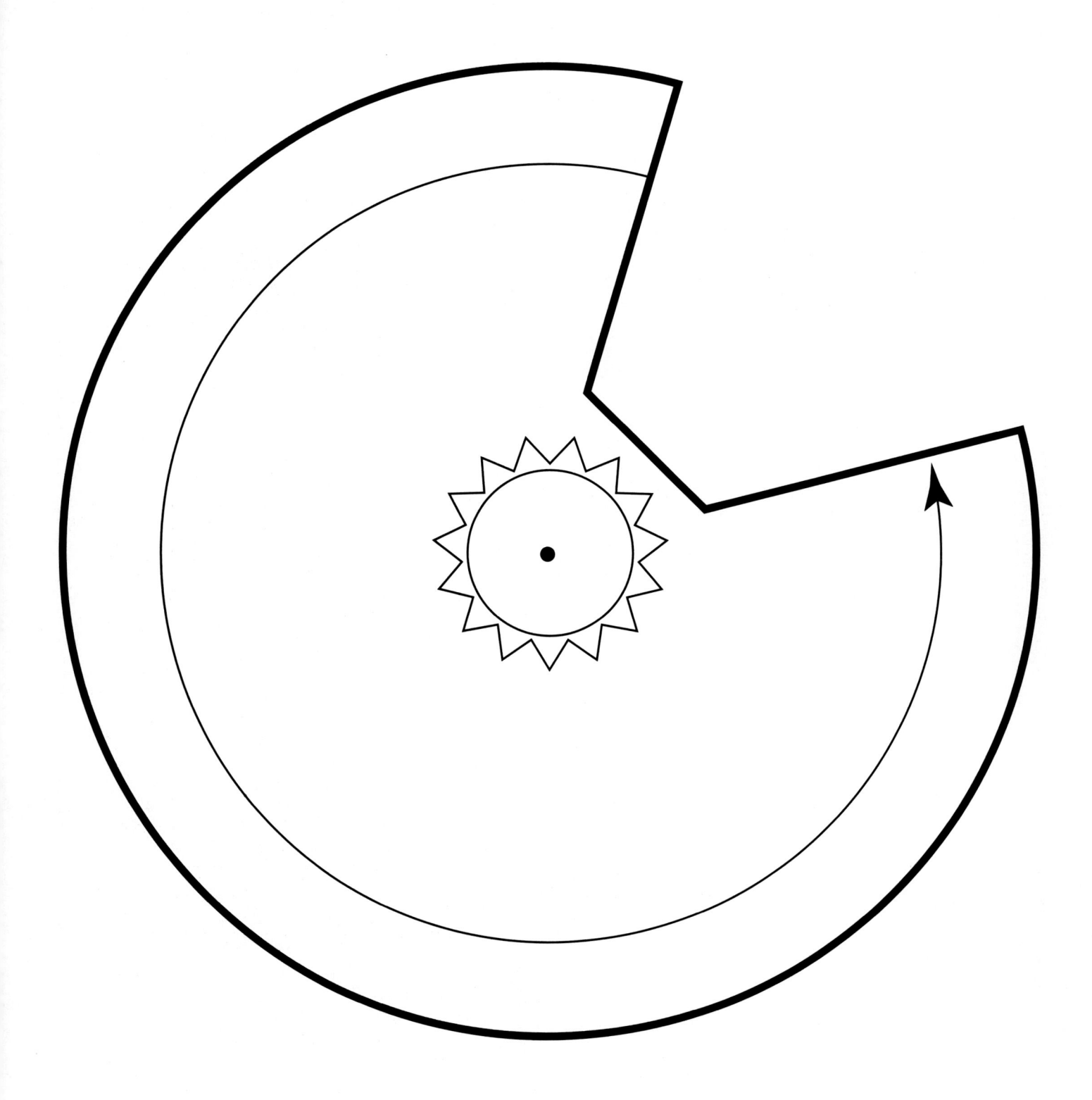

Model Piece: Top
Use with the activity on page 107.

108 ©The Mailbox® • *Science in a Box* • TEC60894

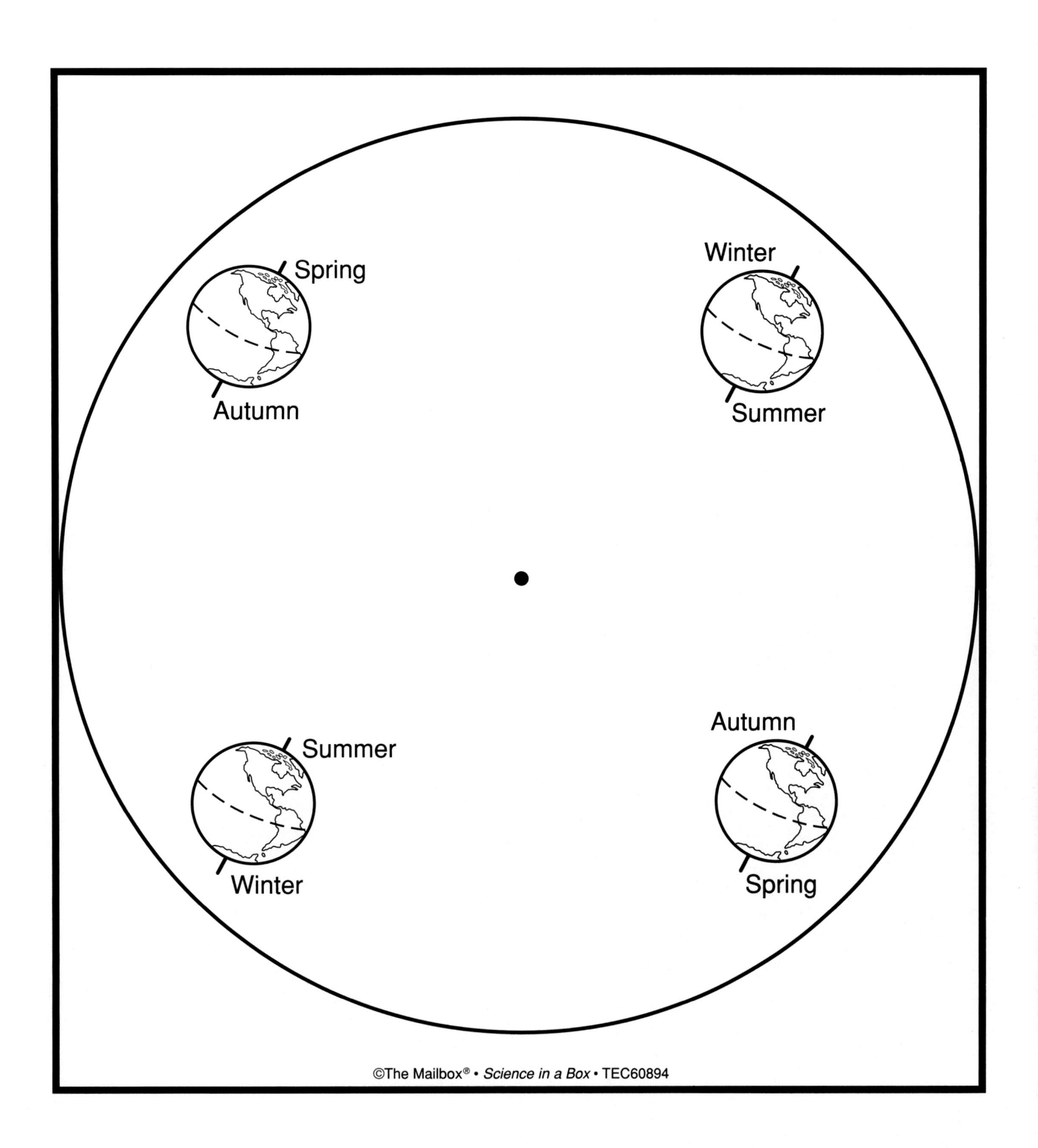

Model Piece: Base
Use with the activity on page 107.

©The Mailbox® • *Science in a Box* • TEC60894

Booklet Pages
Use with the center idea on page 107.

1. When it is autumn in the Southern Hemisphere, it is spring in the Northern Hemisphere.

2. When it is winter in the Southern Hemisphere, it is summer in the Northern Hemisphere.

3. When it is spring in the Southern Hemisphere, it is autumn in the Northern Hemisphere.

4. When it is summer in the Southern Hemisphere, it is winter in the Northern Hemisphere.

Booklet Pieces
Use with the center idea on page 107.

Booklet Page and Cover
Use with the center idea on page 107.

_____ is my favorite season.

5

Four Seasons Make a Year

Name _____

©The Mailbox® • Science in a Box • TEC60894

Answer Key

When it is autumn in the Southern Hemisphere, it is spring in the Northern Hemisphere.
1

When it is spring in the Southern Hemisphere, it is autumn in the Northern Hemisphere.
3

When it is winter in the Southern Hemisphere, it is summer in the Northern Hemisphere.
2

When it is summer in the Southern Hemisphere, it is winter in the Northern Hemisphere.
4

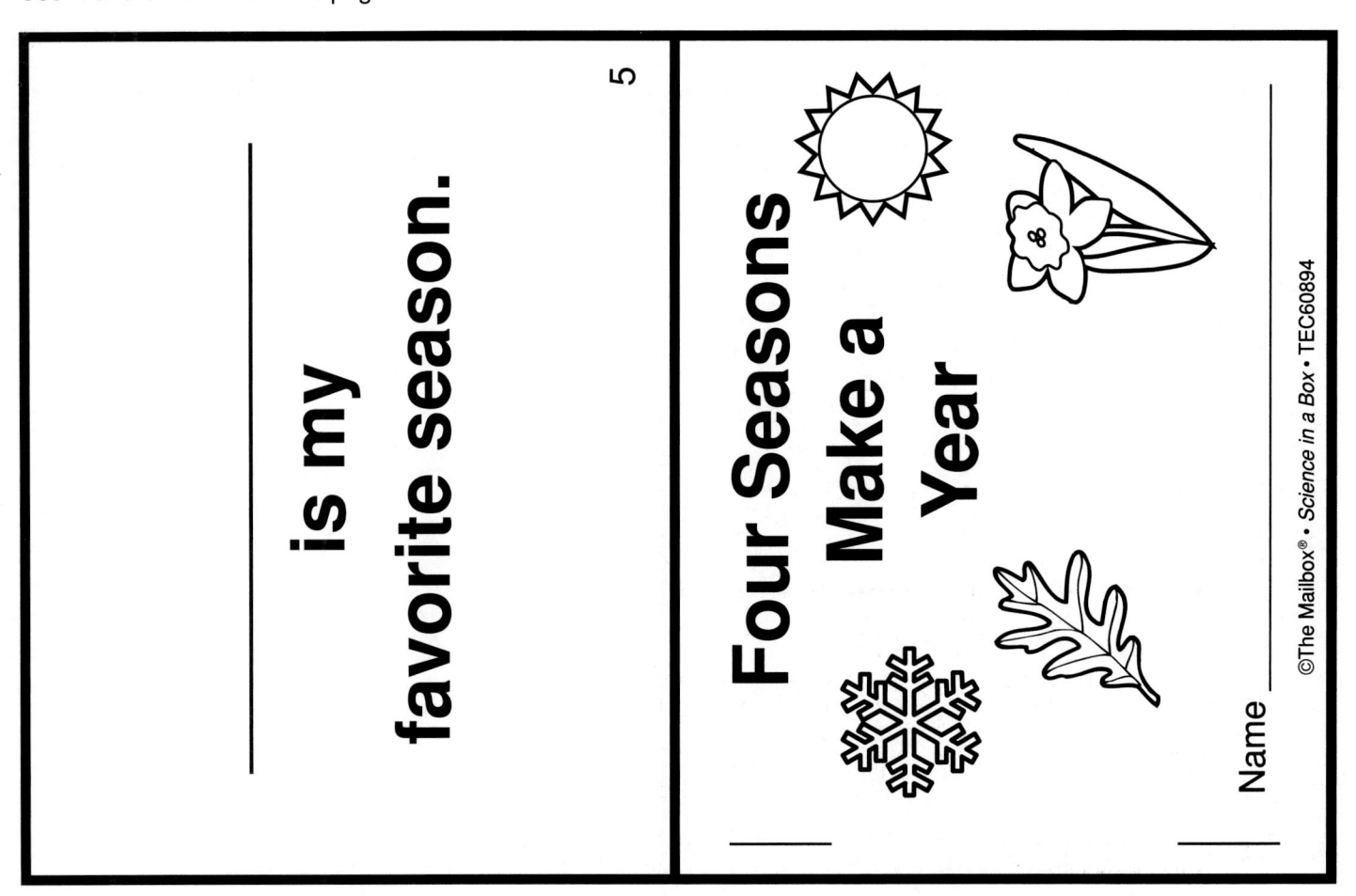

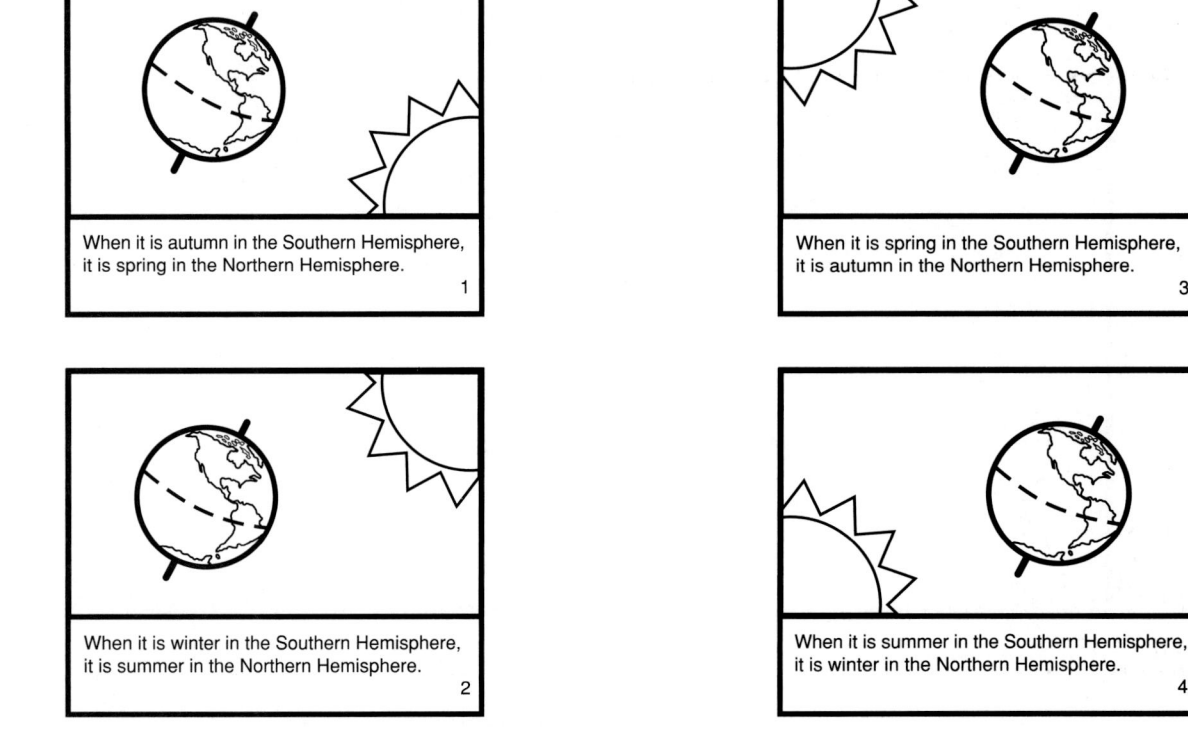

112

SOIL

Objective: to observe and describe different types of soil

Materials: pages 113 and 115, magnifying glasses, drawing paper, jar of soil with a lid for each small group, water

Teacher preparation:
1. Follow the directions on page 3 to assemble the box.
2. Laminate this page and page 115 for durability.
3. Place the laminated pages, magnifying glasses, and drawing paper in the box. If desired, also store the remaining pages from this unit in the box.

Activity

Begin the activity by displaying the diagram on page 115. Point out each layer of soil to students and explain to them that topsoil is the richest layer of soil. Next, divide the class into small groups. Give each group a jar of soil and a magnifying glass. Then give each group member a sheet of drawing paper. Direct the child to fold the paper in half horizontally and then unfold it. On one half of the paper, have him draw the jar of soil. Next, have one member of the group add water to the jar until it is filled almost to the top. Help him screw the lid tightly onto the jar. Then have group members take turns shaking the jar. After each child has had a turn, have the group gently set its jar on a flat surface and leave it for five minutes. Instruct each group member to draw what he sees on the second half of his paper without touching the jar, using his magnifying glass as necessary. Lead groups in a discussion of their findings.

This Is Why

Groups should find that the soil in their jars has settled into layers. The soil settles according to the size and shape of its particles.

Center Idea

Laminate page 117 for durability. Collect small samples of soil, sand, and clay and put each one into a different resealable bag. Label the bags "1," "2," and "3." Place the bags in a center along with page 117 and copies of page 114. A child uses page 117 and the samples to complete page 114.

©The Mailbox® • Science in a Box • TEC60894

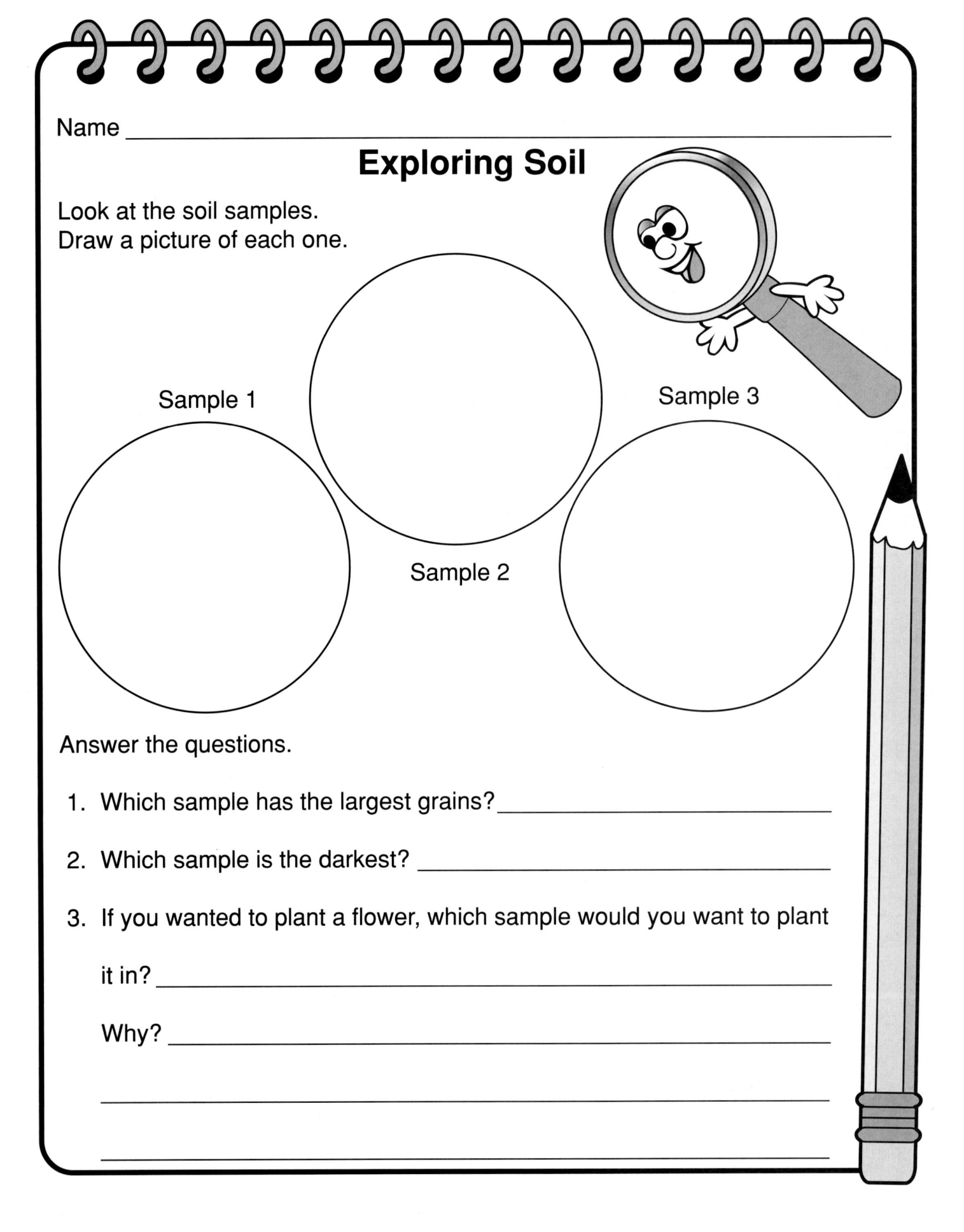

Name _____

Exploring Soil

Look at the soil samples.
Draw a picture of each one.

Sample 1

Sample 2

Sample 3

Answer the questions.

1. Which sample has the largest grains? _____

2. Which sample is the darkest? _____

3. If you wanted to plant a flower, which sample would you want to plant it in? _____

   Why? _____
   _____

Where Is Soil?

← Topsoil

← Subsoil

← Bedrock

Types of Soil

Clay

Sand

Loam

Ask yourself

- What color is the sample?
- What size are the grains?
- What is the texture of the sample?
  —Is it sticky?
  —Is it dry?
- Are there pieces of plants or small bugs in the sample?

ROCKS

Objective: to understand that rocks can be classified by the ways they are formed

Materials: pages 119–122

Teacher preparation:
1. Follow the directions on page 3 to assemble the box.
2. Laminate this page and the cards on page 121 for durability. Cut apart the rock cards.
3. Make a class supply of page 120.
4. Place this page, the rock cards, and the copies inside the box. If desired, also store the remaining pages from this unit in the box.

Activity

Begin the activity by telling students that rocks can be formed in three ways. Give each child a copy of page 120 and have him cut the fact cards apart. Read and discuss the fact cards with the class. Next, display one rock card. Read aloud the information on the back of the card (without revealing the rock type). Instruct each student to study his fact cards, determine which type of rock is on the displayed card, and hold up the appropriate fact card. Lead students in a discussion of their choices before revealing the correct type. Repeat the activity with each of the remaining rock cards. Then instruct students to store their fact cards in a safe place to use as a reference or to help them complete the center idea suggested on this page.

Igneous

Rock that has melted, cooled, and hardened

This will help you remember! Think of hard candy that changes from a liquid to a solid when it cools down.

This Is Why

Rocks can by classified by the way they are formed. There are three ways that rocks are formed. Igneous rocks form from heated magma deep inside the earth that is cooled when it reaches the earth's surface. Sedimentary rocks are formed from materials that settle into layers. Metamorphic rocks are formed when rock is changed by heat or movements in the earth or both.

Center Idea

Place a class supply of page 123 and the top half of page 124, several rock reference books, and crayons at a center. A child refers to his fact cards as he writes the word *igneous*, *metamorphic*, or *sedimentary* on each blank line. Next, he looks up each rock in the reference books and draws it on the appropriate booklet page. Then he cuts out the pages, stacks them, and staples them together for a handy rock reference.

©The Mailbox® • Science in a Box • TEC60894

Fact Cards
Use with the activity on page 119.

Igneous

rock that has melted, cooled, and hardened

This will help you remember!
Think of hard candy that changes from a liquid to a solid when it cools down.

Sedimentary

rock that forms from material that settles into layers and is squeezed and hardened into rock

This will help you remember!
The word *sediment* is in the word *sedimentary*. Sediment is small pieces of rock and soil that settle in layers.

Metamorphic

rock that has been changed by heat and pressure

This will help you remember!
Metamorphic rock was once igneous rock, sedimentary rock, or another kind of metamorphic rock before it was changed to a new kind of metamorphic rock.

Rock Cards
Use with the activity on page 119.

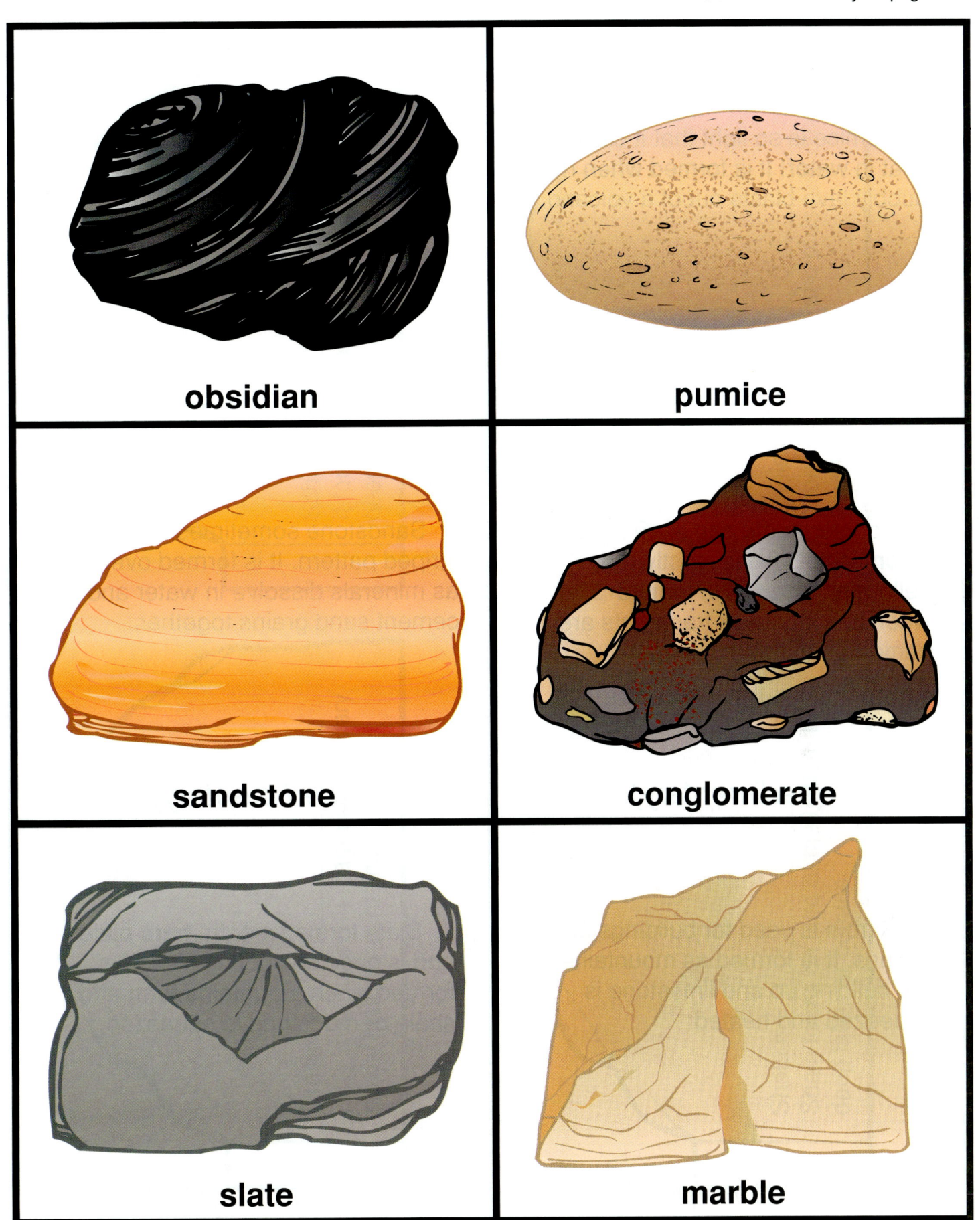

obsidian	pumice
sandstone	conglomerate
slate	marble

126

©The Mailbox® • *Science in a Box* • TEC60894

Name _____

Environmental Actions

Safe Actions	Harmful Actions

©The Mailbox® • Science in a Box • TEC60894

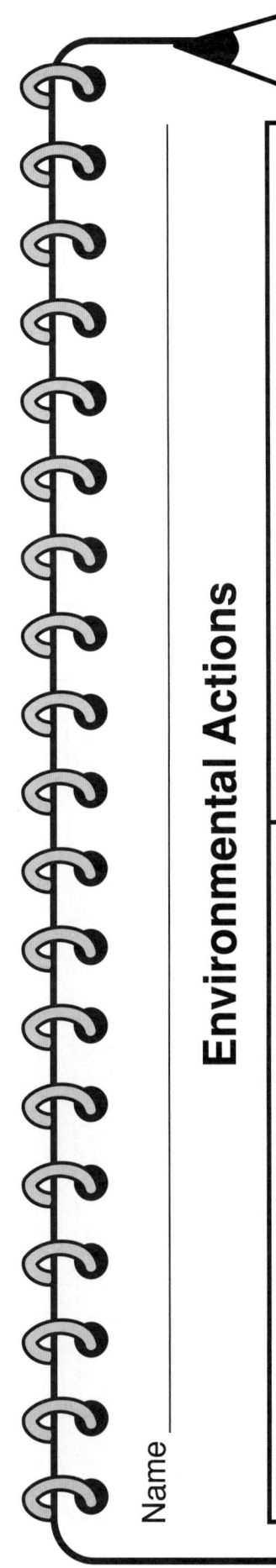

Note to the teacher: Use with the activity on page 125.

Name ____________________

Problem and Solution

Problem: ____________________

My solution: ____________________

____

____

____

 Note to the teacher: Use with the center idea on page 125.

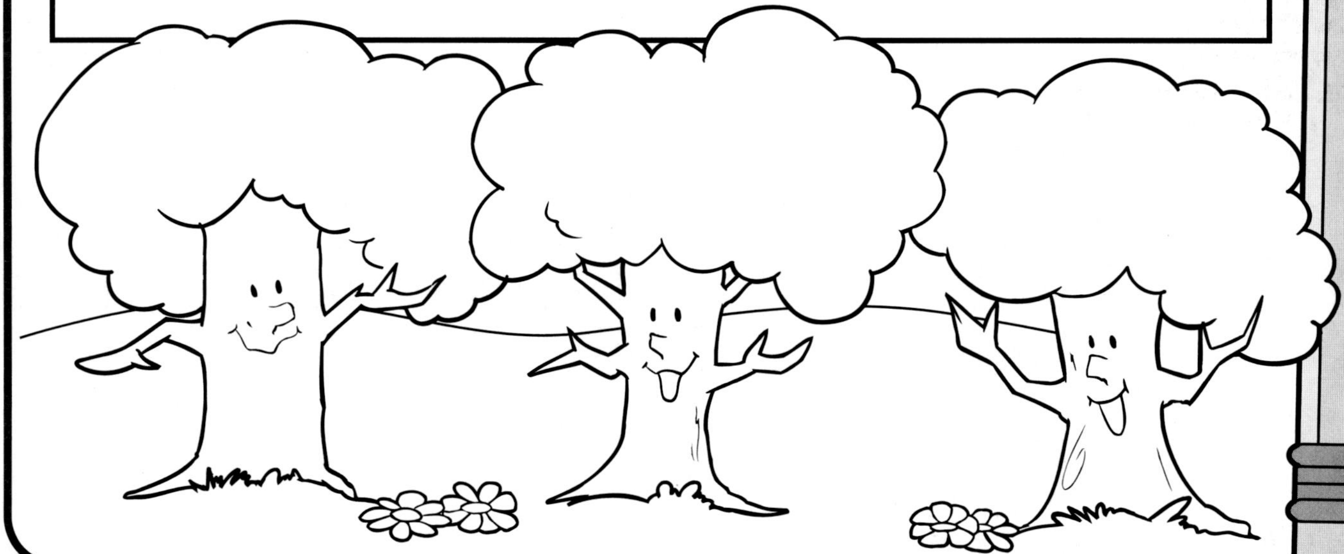

Problem Cards
Use with the center idea on page 125.

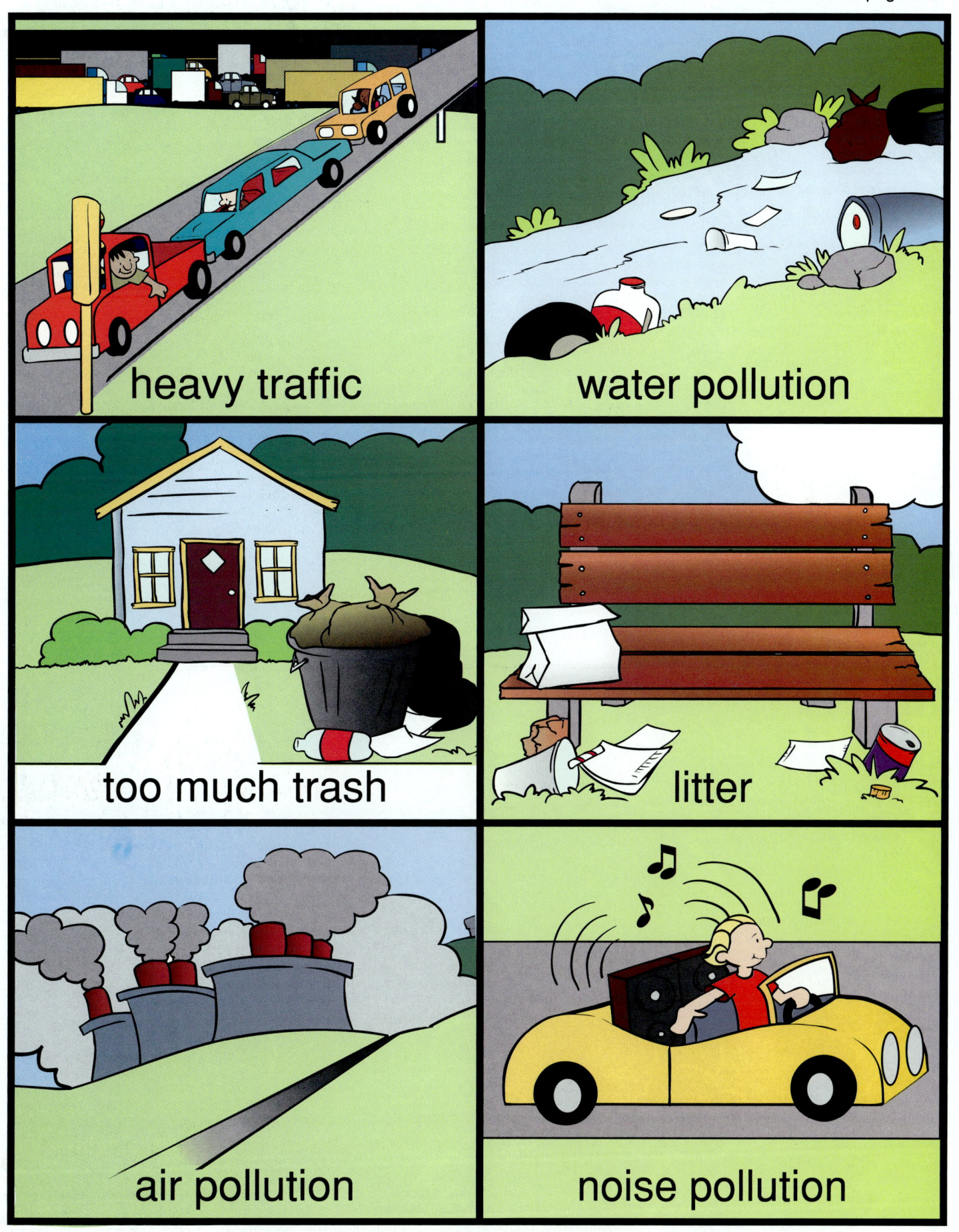

heavy traffic

water pollution

too much trash

litter

air pollution

noise pollution

©The Mailbox® • *Science in a Box* • TEC60894

POSITION OF THE SUN

Objective: to understand that the position of the sun changes in the sky as the earth rotates

Materials: pages 131, 133–135, yellow construction paper

Teacher preparation:
1. Follow the directions on page 3 to assemble the box.
2. Laminate this page and the sun cards on page 135 for durability. Cut apart the cards.
3. Make one copy of pages 133 and 134 for each student.
4. Place this page, the sun cards, and the copies inside the box.

Activity

Begin by asking students whether the sun moves across the sky throughout the day. Allow students to respond; then point out that it's actually the earth's movement that makes the sun appear in different places in the sky throughout the day. Explain that the earth spins, or rotates, like a top. One full rotation takes one full day (24 hours). As the earth turns toward the sun, it is day in that part of the earth. As the earth turns away from the sun, it is night in that part of the earth. Share the sun cards to illustrate how the sun appears in the sky at different times of the day. Tape each card to the board in chronological order. Then guide each student to make a sun booklet.

First, give each student one copy of pages 133 and 134 and a half sheet of construction paper. To make her booklet, each student illustrates an activity for each time of day specified on the booklet pages. Then she adds a sun in the correct position for each picture. Next, she cuts out the booklet pages and then traces booklet page 1 onto yellow construction paper, making a booklet cover. Finally, she personalizes and cuts out the cover and then staples it atop her booklet pages.

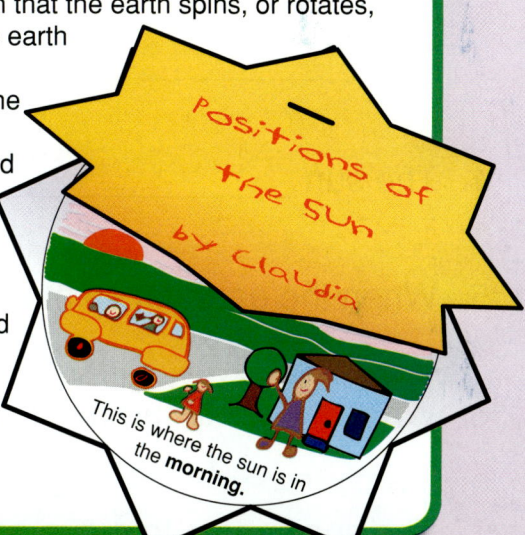

This Is Why

The earth spins, or rotates, in space. It takes 24 hours to make one rotation. As the earth rotates, the sun seems to move across the sky. However, the earth's movement is what causes day and night.

Center Idea

Place the sun cards and a class supply of page 132 at a center. A student arranges the cards chronologically to show the position of the sun throughout the day. Then he completes a copy of page 132.

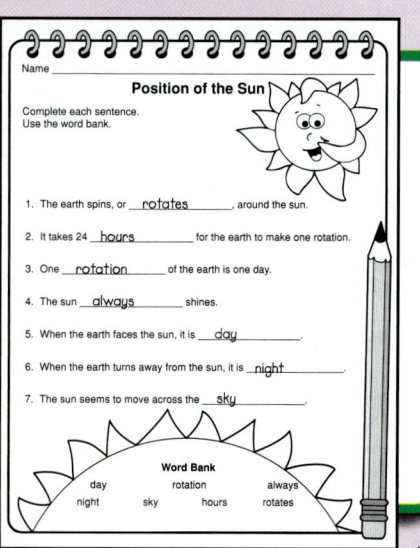

©The Mailbox® • *Science in a Box* • TEC60894

Name _____

Position of the Sun

Complete each sentence.
Use the word bank.

1. The earth spins, or _____, around the sun.

2. It takes 24 _____ for the earth to make one rotation.

3. One _____ of the earth is one day.

4. The sun _____ shines.

5. When the earth faces the sun, it is _____.

6. When the earth turns away from the sun, it is _____.

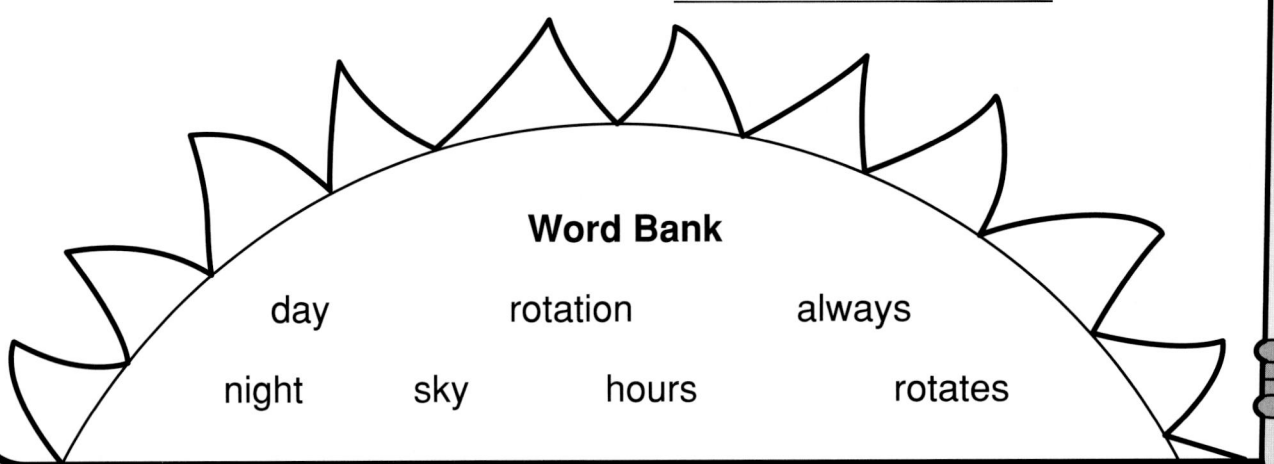

7. The sun seems to move across the _____.

Word Bank

day	rotation	always
night	sky	hours	rotates

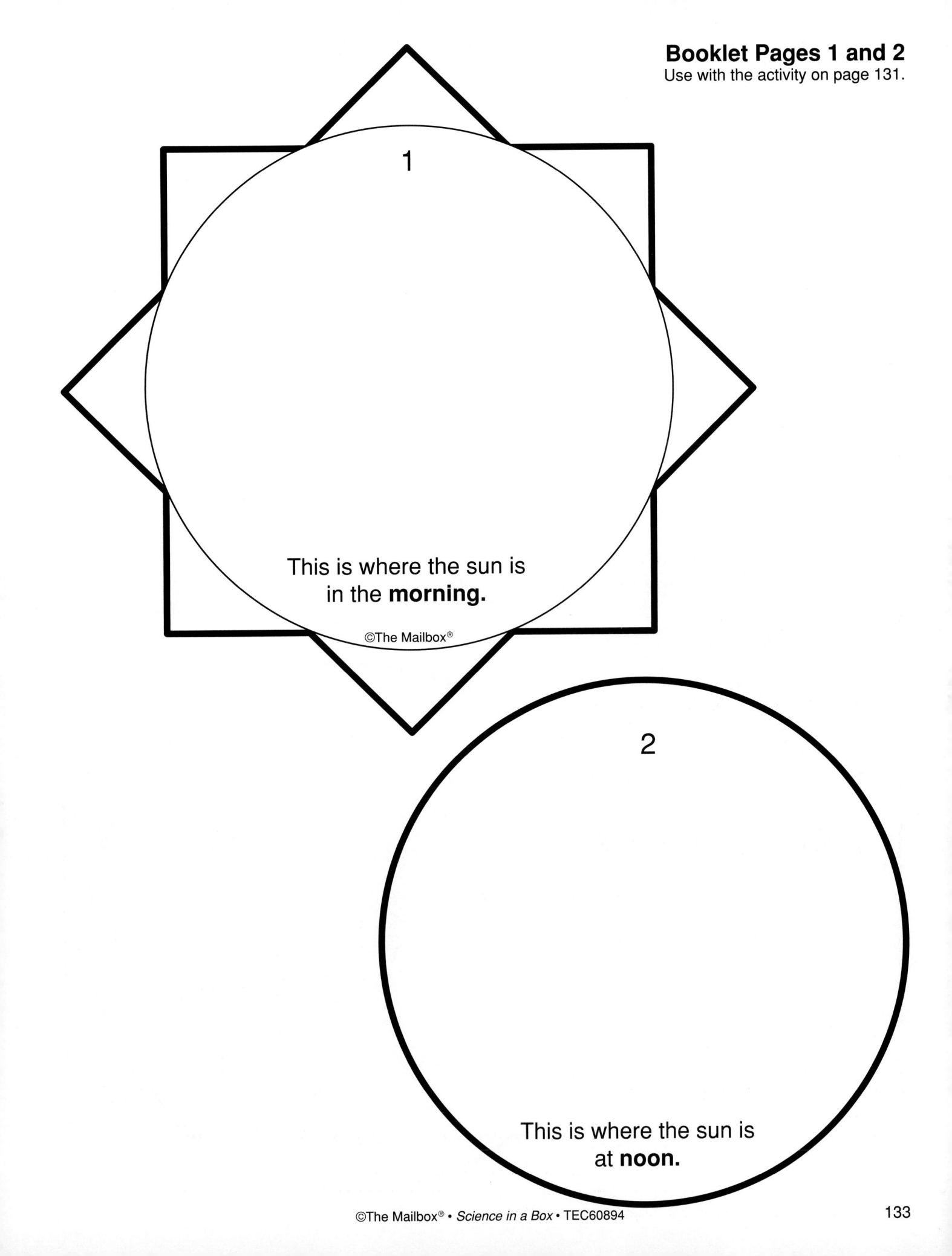

Booklet Pages 1 and 2
Use with the activity on page 131.

1

This is where the sun is in the **morning.**

©The Mailbox®

2

This is where the sun is at **noon.**

©The Mailbox® • *Science in a Box* • TEC60894

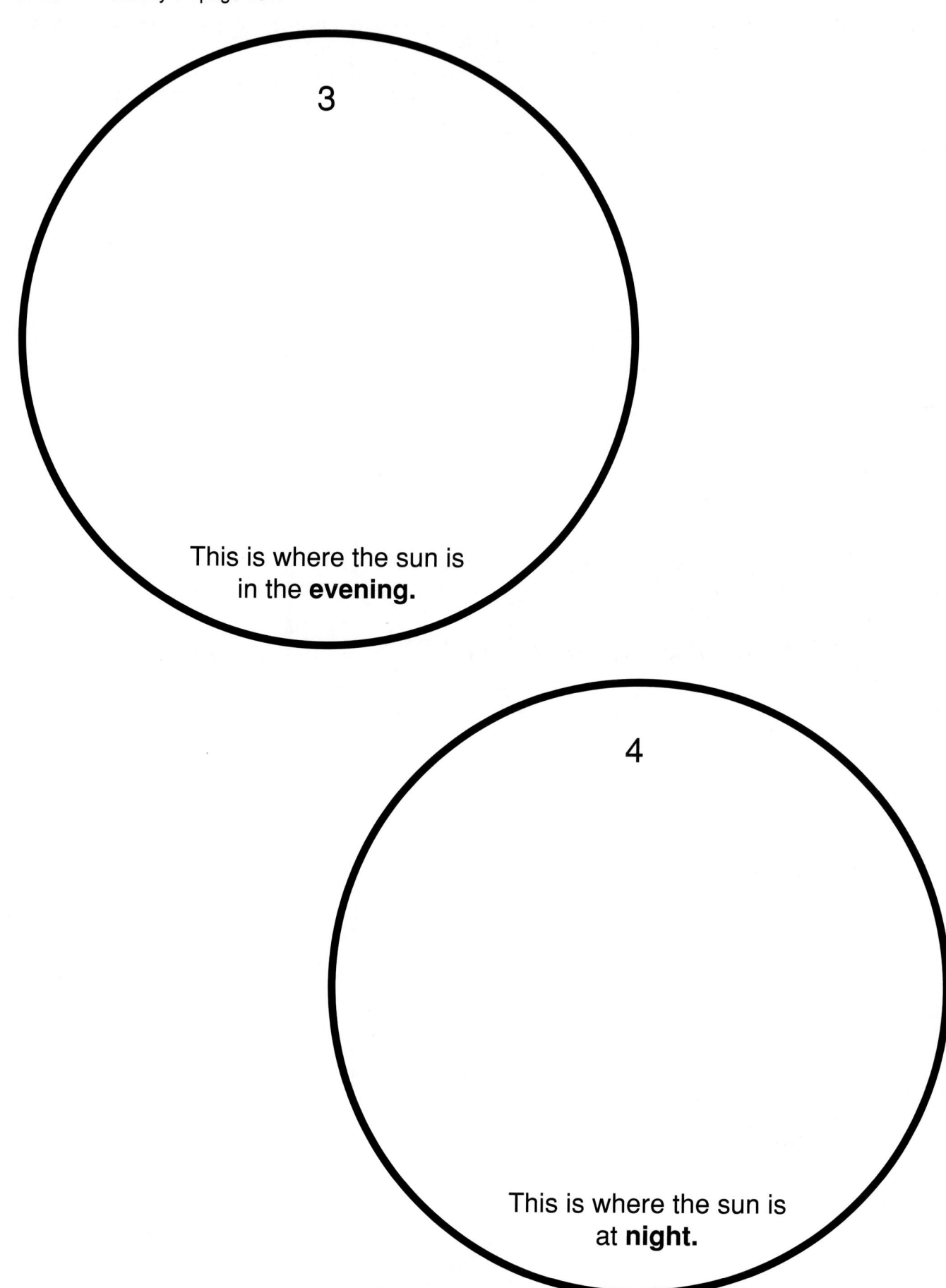

Booklet Pages 3 and 4
Use with the activity on page 131.

3

This is where the sun is in the **evening.**

4

This is where the sun is at **night.**

134 ©The Mailbox® • *Science in a Box* • TEC60894

Sun Cards

Use with the activity and center idea on page 131.

noon

night

morning

evening

PHASES OF THE MOON

Objective: to understand the lunar cycle

Materials: pages 137–139

Teacher preparation:
1. Follow the directions on page 3 to assemble the box.
2. Laminate this page and the moon cards on page 139 for durability. Cut apart the cards.
3. Make a class supply of page 138.
4. Place this page, the moon cards, and the copies in the box. If desired, also store the remaining pages from this unit in the box.

Activity

Begin the activity by displaying the moon cards in random order. Discuss the cards with students as you place each card on the chalk tray. Next, point out the new moon card and tape it to the board. Have students help you identify the moon phase that comes next in the lunar cycle *(waxing crescent)* and tape it to the board next to the first card. Continue in this manner until all of the cards are taped in order *(new moon, waxing crescent, first quarter, waxing gibbous, full moon, waning gibbous, third quarter, waning crescent)*. Explain to students that after the last phase, the cycle begins again. Further explain that each cycle takes 29½ days. For individual practice, have each child complete a copy of page 138.

This Is Why

The moon orbits Earth. As the moon moves around Earth, it reflects the sun's light. As different parts of the moon reflect light, those parts become visible to Earth at different times in the lunar cycle.

Center Idea

Make one copy of pages 141 and 142. Laminate each page and then cut out the gameboards, game cards, and game directions. Place the game cards in a resealable plastic bag and place them at a center along with the gameboards and directions. Invite two students at a time to visit the center and follow the directions to play the game.

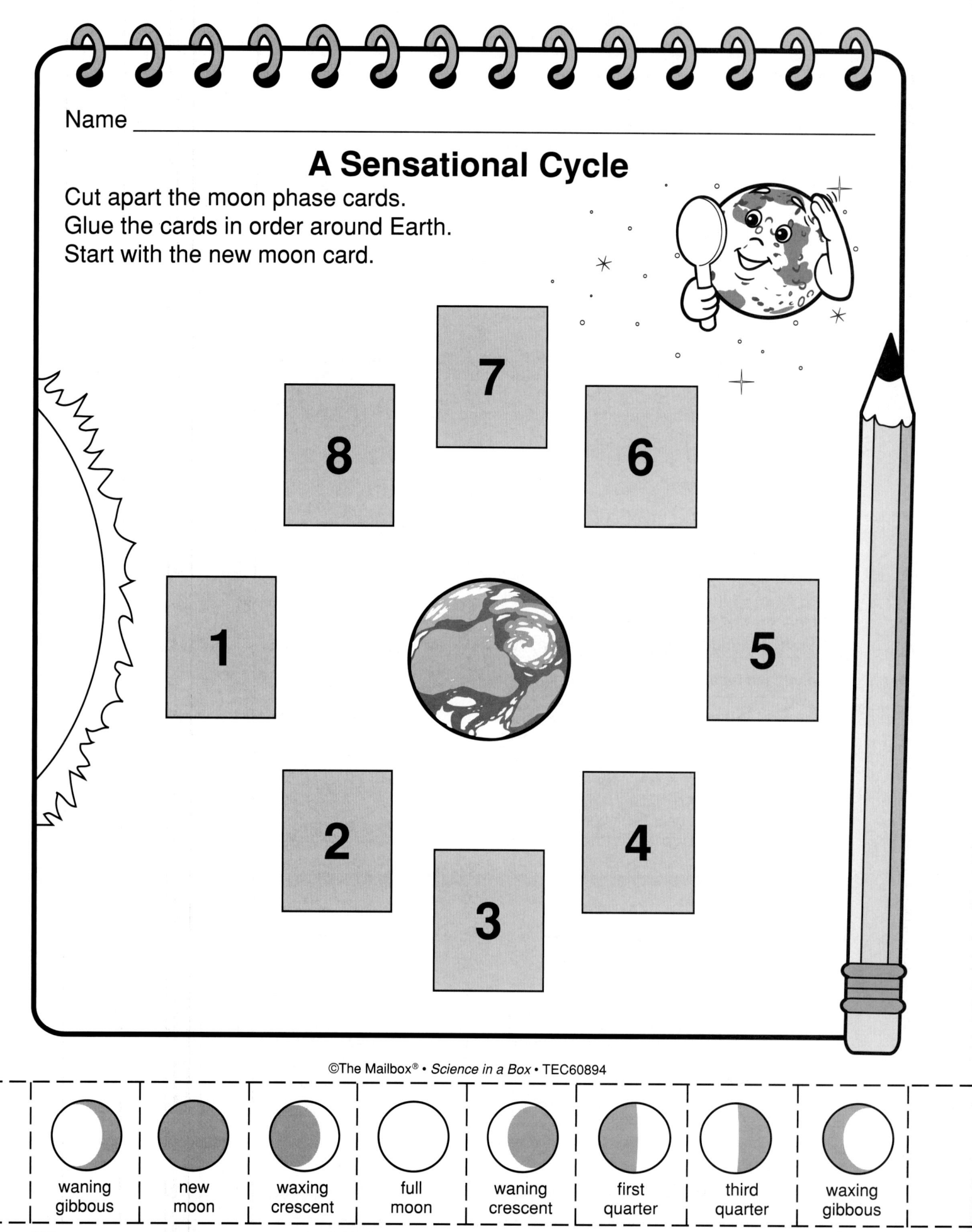

Name _____

A Sensational Cycle

Cut apart the moon phase cards.
Glue the cards in order around Earth.
Start with the new moon card.

		7		
	8		6	
1		🌍		5
	2		4	
		3		

©The Mailbox® • Science in a Box • TEC60894

| waning gibbous | new moon | waxing crescent | full moon | waning crescent | first quarter | third quarter | waxing gibbous |

138 **Note to the teacher:** Use with the activity on page 137.

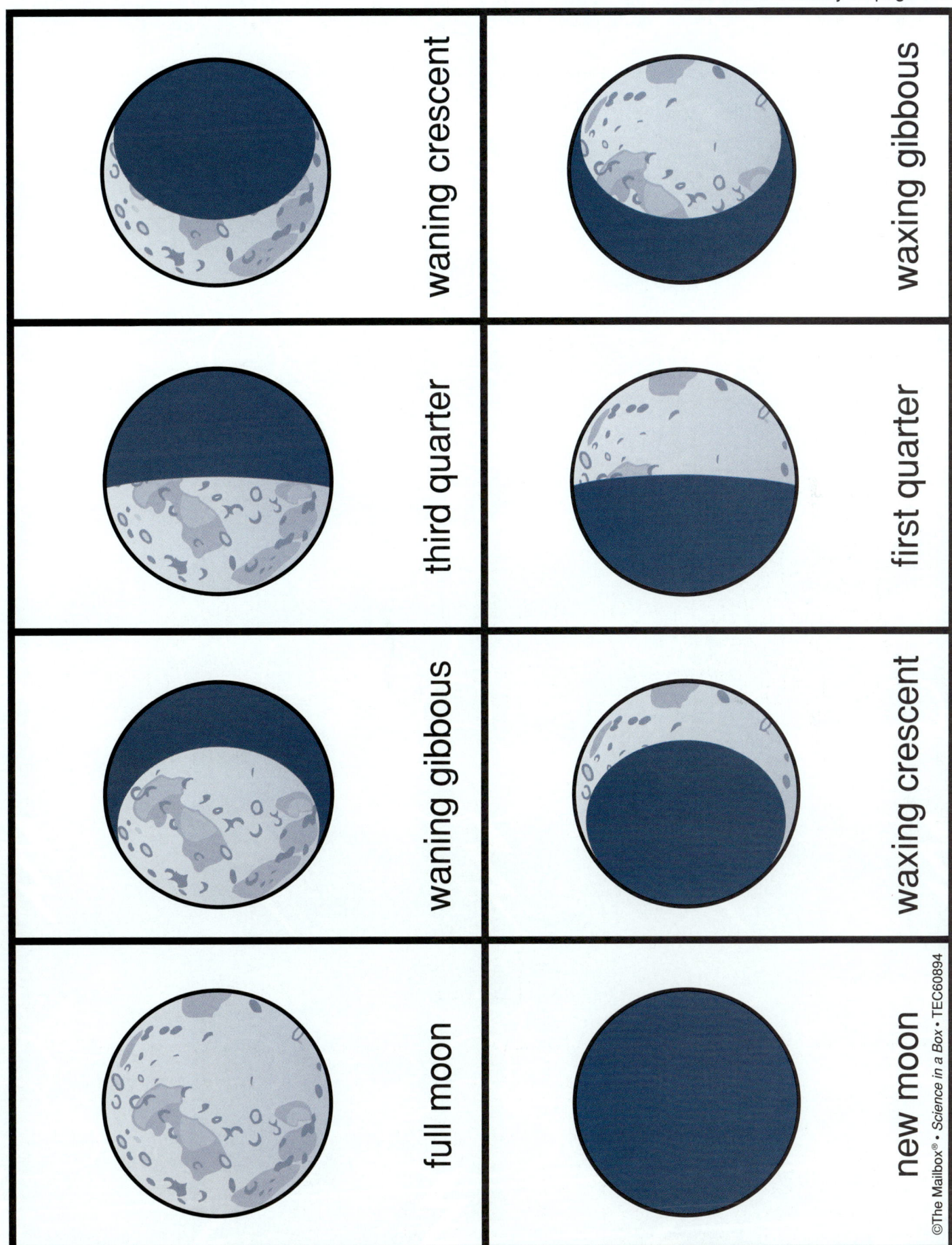

Moon Cards
Use with the activity on page 137.

waning crescent	waxing gibbous
third quarter	first quarter
waning gibbous	waxing crescent
full moon	new moon

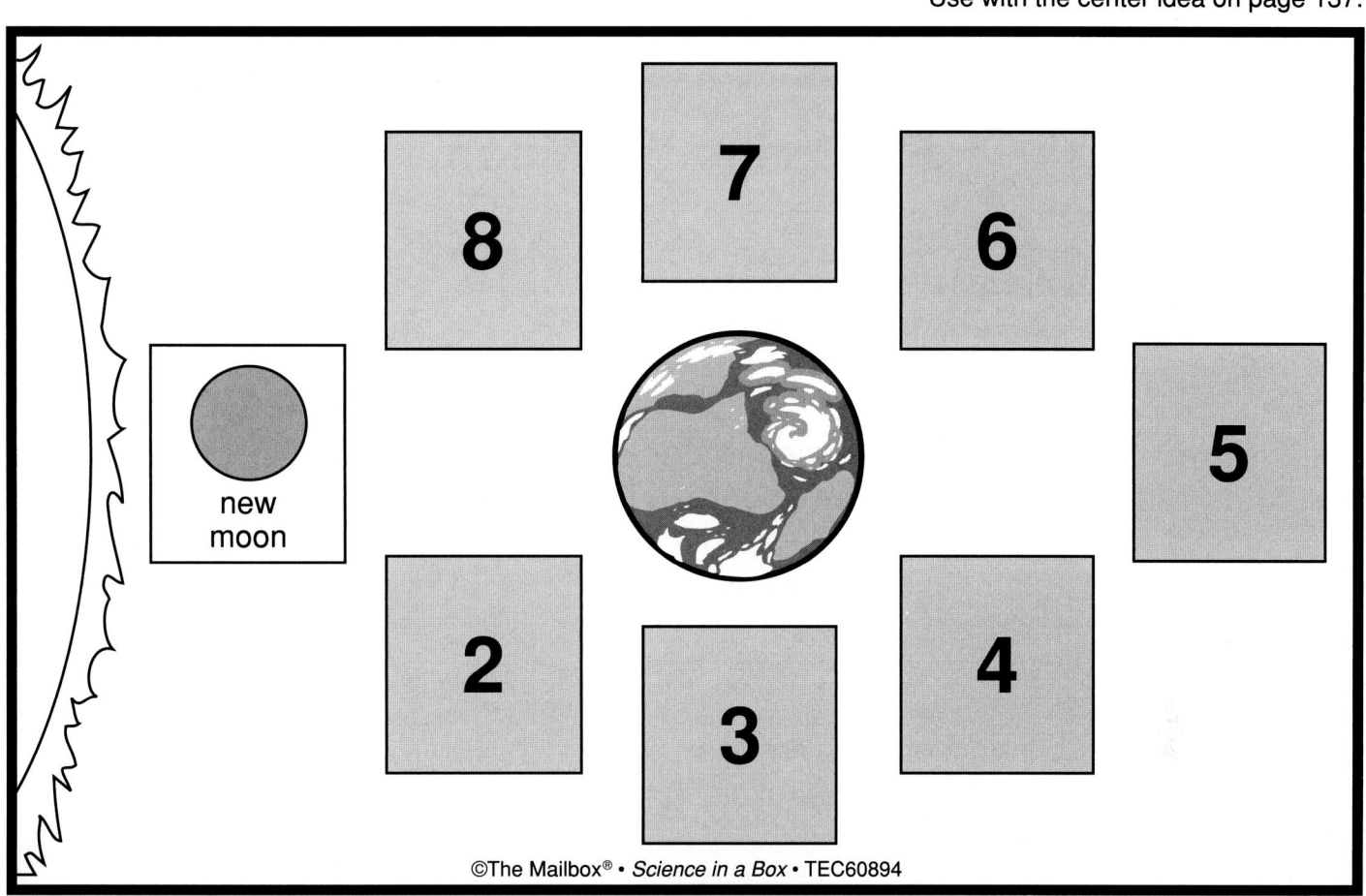

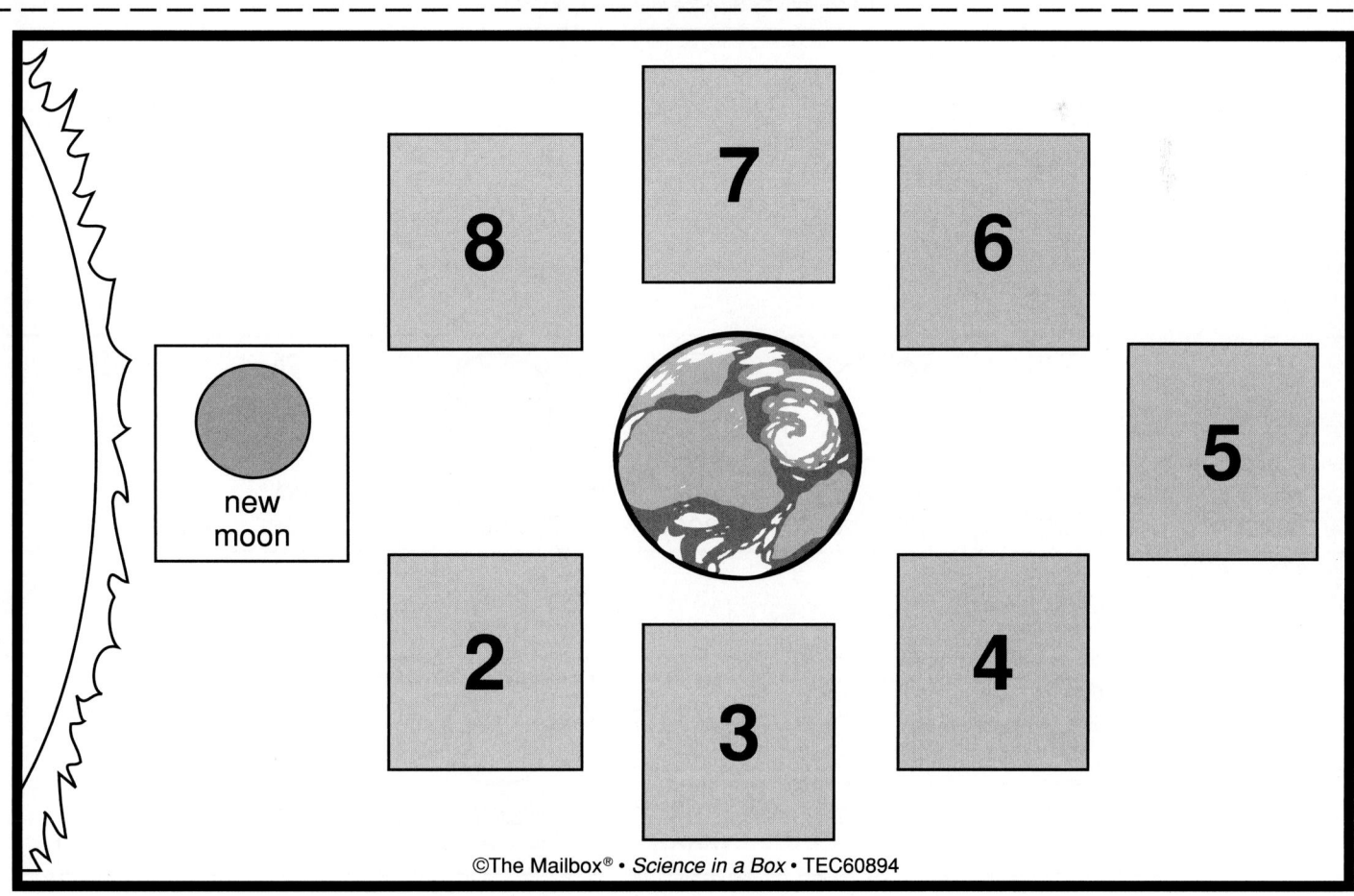

Gameboards
Use with the center idea on page 137.

new moon

8	7	6
2	3	4

5

©The Mailbox® • *Science in a Box* • TEC60894

---

new moon

8	7	6
2	3	4

5

©The Mailbox® • *Science in a Box* • TEC60894

Game Cards and Directions

Use with the center idea on page 137.

Phases of the Moon

To Play:

1. Each player places a gameboard in front of him. One player shuffles the cards and lays them facedown.

2. Player 1 draws a card and places it on the correct space of his gameboard. If the space is already covered, he returns that card to the bottom of the card pile.

3. Player 2 takes a turn in the same manner.

4. Play continues until one player has correctly completed his lunar cycle.

©The Mailbox® • *Science in a Box* • TEC60894

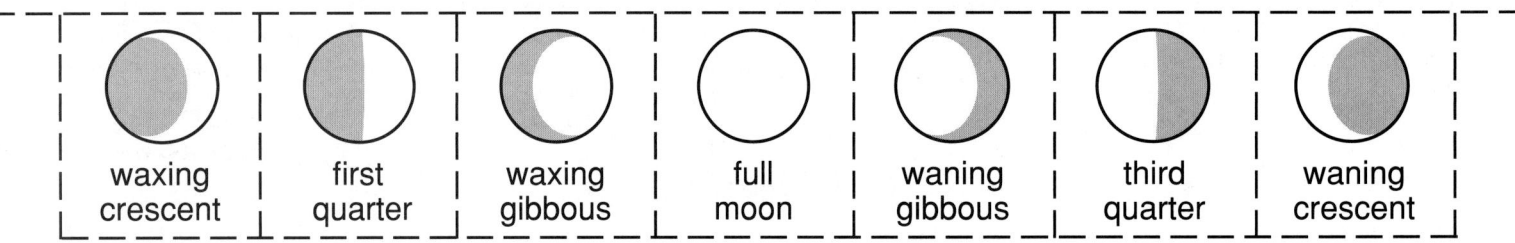

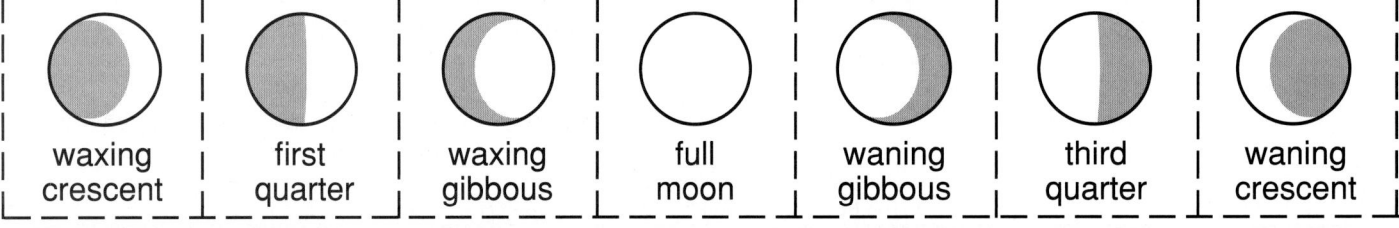

| waxing crescent | first quarter | waxing gibbous | full moon | waning gibbous | third quarter | waning crescent |

| waxing crescent | first quarter | waxing gibbous | full moon | waning gibbous | third quarter | waning crescent |

SOLIDS AND LIQUIDS

Objective: to understand that a solid has a definite shape and a liquid does not

Materials: pages 143 and 145, 1 c. water (or other liquid), 2 clear glass jars, board eraser

Teacher preparation:
1. Follow the directions on page 3 to assemble the box.
2. Laminate this page for durability.
3. Make one copy of the recording sheet at the bottom of page 145 for each small group of students.
4. Place this page, the glass jars, the board eraser, and the copies inside the box. If desired, also store the remaining page from this unit in the box.

Activity

Begin by pouring the water into one of the glass jars. Next, drop the board eraser in the other jar. Ask students to explain the differences in the contents of the two jars. After several responses, explain that the water, which is a liquid, took the shape of the jar. The eraser, which is a solid, did not. Then ask what will happen if you dump the contents of both jars onto the floor. Point out that the water does not have a definite shape, so it will spread out on the floor. However, the eraser will hold its shape and won't look any different.

Next, give each small group of students a copy of the recording sheet (page 145). Challenge each group to fill the chart with examples of solids and liquids that start with the letters listed. Explain that the group will receive one point for each correct response and an extra point if the item was not listed by any other group. After a predetermined amount of time, allow groups to share their answers and award points as described. The group with the most points wins.

Starts With	Solid	Liquid	Points
S	skateboard	soup	3
M	marshmallows	mouthwash	2
B	baseball	blood	2
R	rake	rain	3
D	door	diet soda	4
H	hammer	honey	2

This Is Why

A solid is matter that has a definite shape and a definite volume. A liquid is matter that does not have a definite shape but does have a definite volume. A liquid poured into a container will take the shape of that container.

Center Idea

Cut apart the cards and answer key from page 145 and the cards from page 147. Place the cards, two game markers, the answer key, and the gameboard from page 144 at a center. Invite pairs of students to visit the center and play the game as directed on the gameboard.

Solid or Liquid?

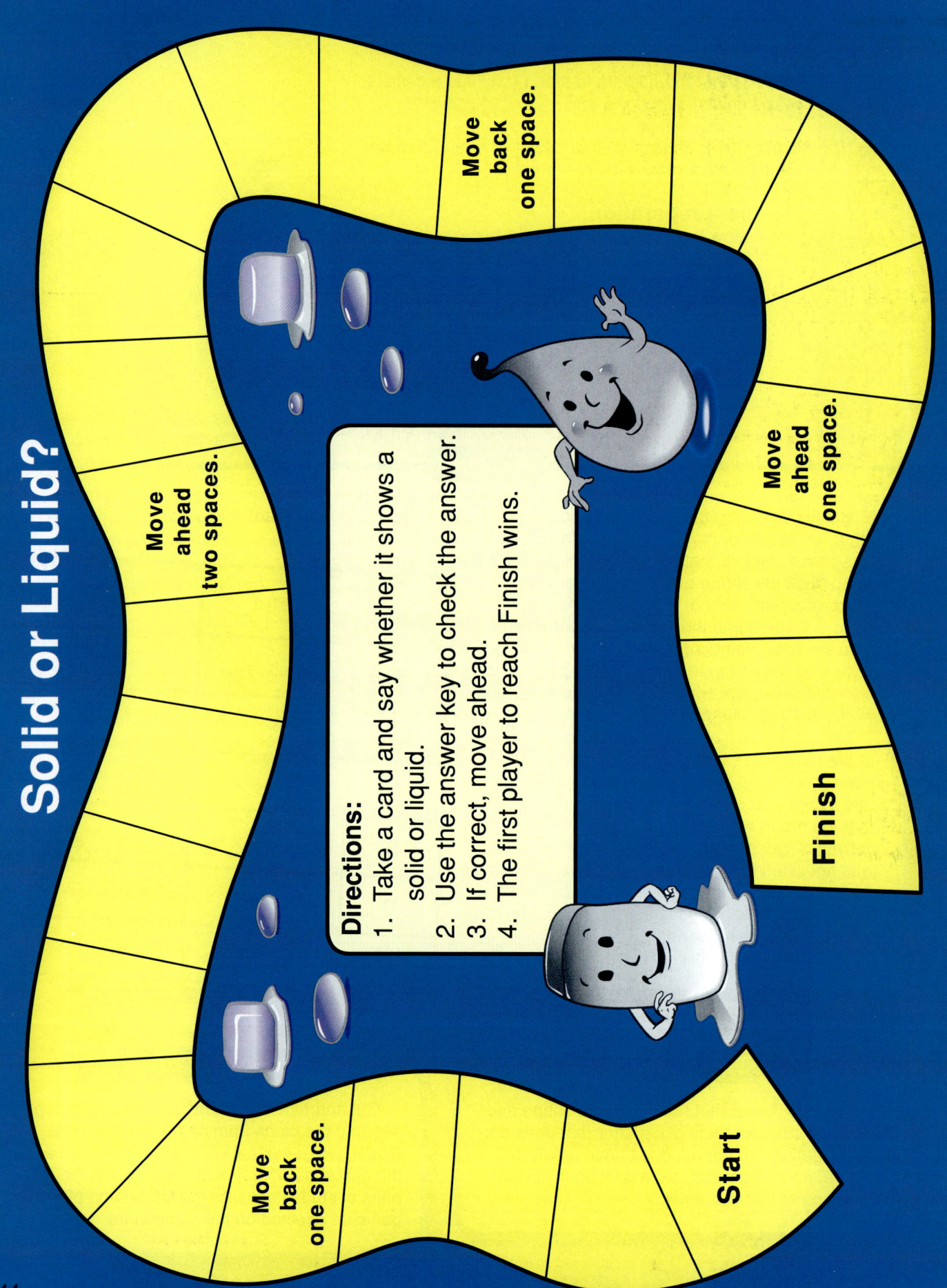

Directions:
1. Take a card and say whether it shows a solid or liquid.
2. Use the answer key to check the answer.
3. If correct, move ahead.
4. The first player to reach Finish wins.

Move back one space.

Move ahead two spaces.

Move ahead one space.

Move back one space.

Finish

Start

Solid and Liquid Cards and Answer Key

Use with the center idea on page 143.

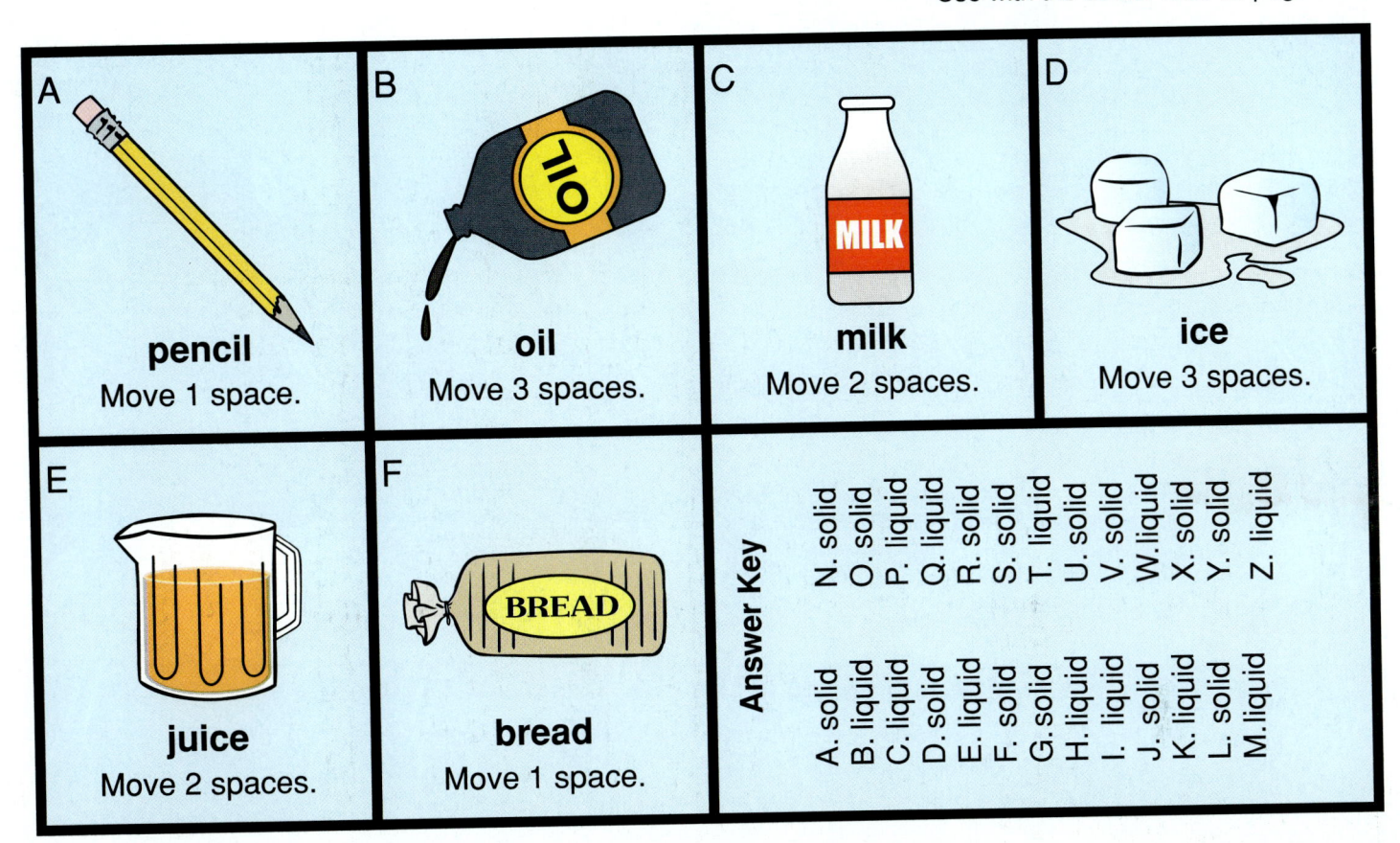

A	B	C	D
pencil Move 1 space.	**oil** Move 3 spaces.	**milk** Move 2 spaces.	**ice** Move 3 spaces.
E	F		
juice Move 2 spaces.	**bread** Move 1 space.		

Answer Key

A. solid
B. liquid
C. liquid
D. solid
E. liquid
F. solid
G. solid
H. liquid
I. liquid
J. solid
K. liquid
L. solid
M. liquid
N. solid
O. solid
P. liquid
Q. liquid
R. solid
S. solid
T. liquid
U. solid
V. solid
W. liquid
X. solid
Y. solid
Z. liquid

Recording Sheet

Use with the activity on page 143.

Starts With	Solid	Liquid	Points
S			
M			
B			
R			
D			
H			

©The Mailbox® • *Science in a Box* • TEC60894

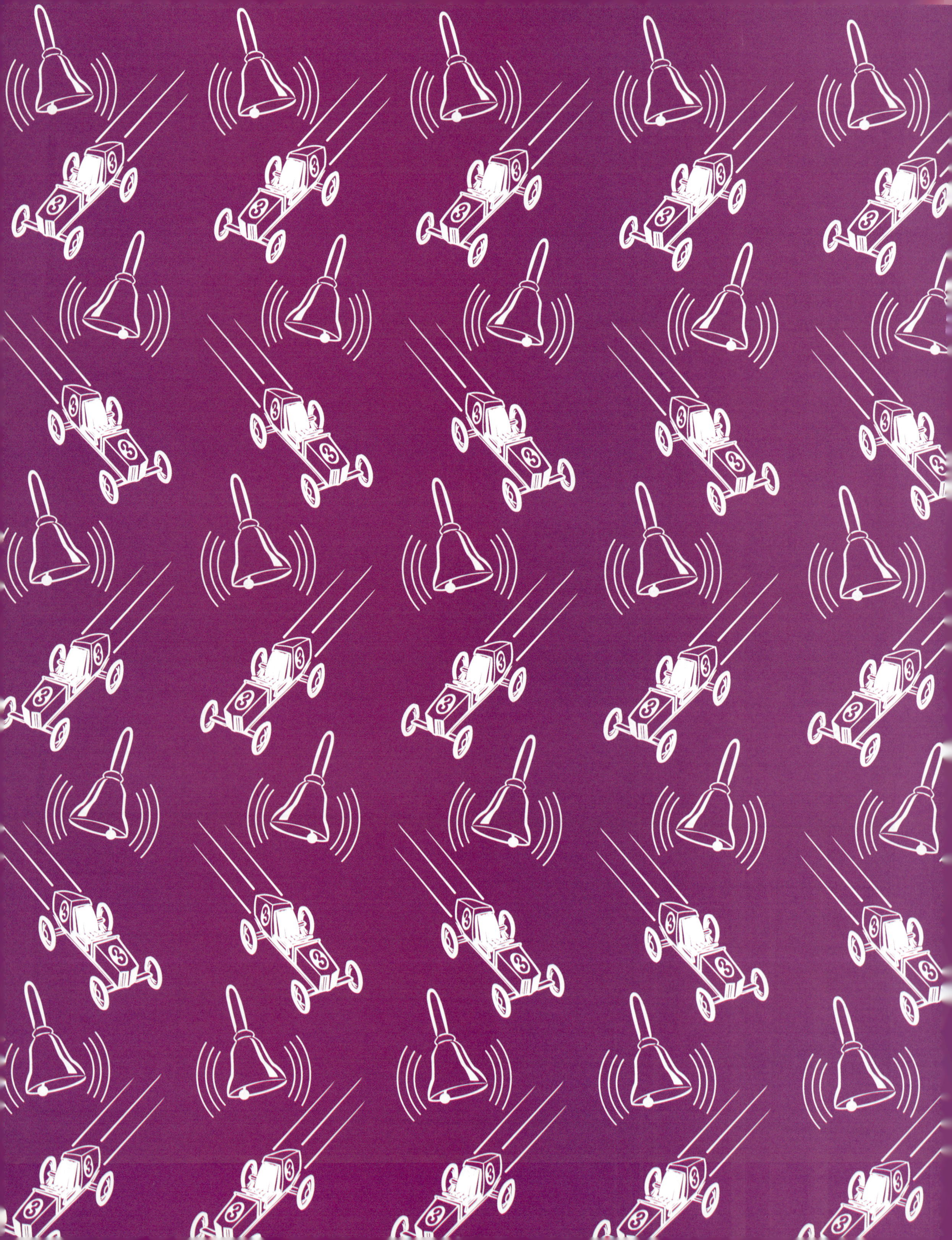

Solid and Liquid Cards
Use with the center idea on page 143.

G	H	I	J
soap Move 2 spaces.	**sunscreen** Move 3 spaces.	**glue** Move 2 spaces.	**paper** Move 1 space.
K	L	M	N
ketchup Move 3 spaces.	**rock** Move 1 space.	**paint** Move 3 spaces.	**book** Move 2 spaces.
O	P	Q	R
eraser Move 1 space.	**shampoo** Move 2 spaces.	**rain** Move 3 spaces.	**glue stick** Move 2 spaces.
S	T	U	V
paper clip Move 1 space.	**chocolate syrup** Move 3 spaces.	**chocolate** Move 2 spaces.	**leaf** Move 2 spaces.
W	X	Y	Z
soda Move 1 space.	**carrot** Move 2 spaces.	**teddy bear** Move 3 spaces.	**soup** Move 1 space.

©The Mailbox® • *Science in a Box* • TEC60894

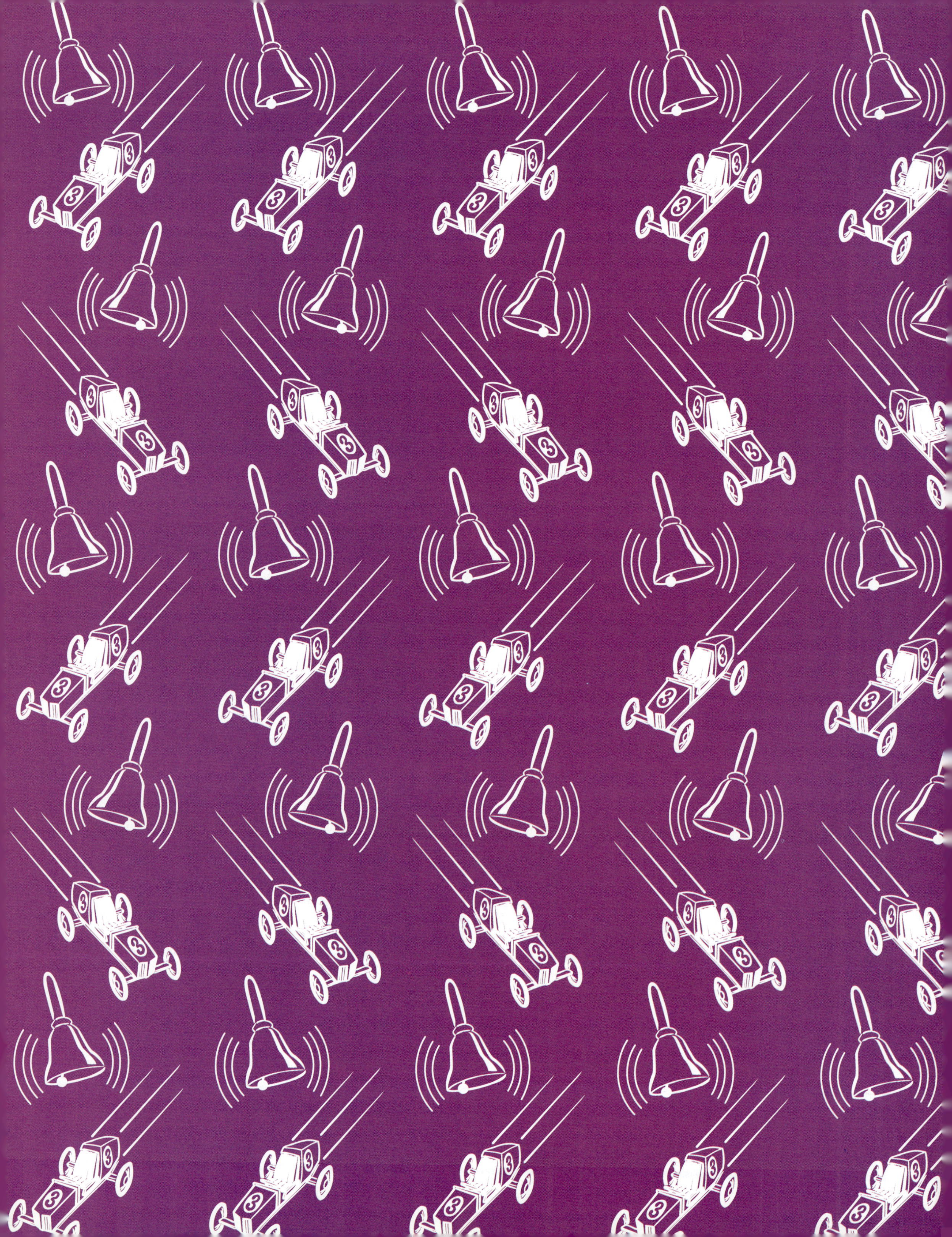

SOUND

Objective: to understand that vibrations cause sound

Materials: pages 149–150

Teacher preparation:
1. Follow the directions on page 3 to assemble the box.
2. Laminate this page for durability.
3. Make a copy of page 150 (omitting the answer key) for each group of students.
4. Place this page and the copies inside the box. If desired, also store the remaining pages from this unit in the box.

Activity

Begin by explaining to students that sounds are made by vibrations that travel in waves to our ears. Ask students to name different musical instruments. List responses on the board. Have students study the list and suggest categories that describe the ways these instruments make sounds *(a surface being struck—percussion instruments, a string being vibrated—stringed instruments, and air being vibrated—wind instruments)*. Next, divide students into groups and give each group a copy of page 150 to cut apart. Instruct group members to discuss how they think the sound of each pictured instrument is made and to sort the cards accordingly into three groups: percussion instruments, stringed instruments, and wind instruments. Then, as a class, discuss which cards belong in each category and allow group members to make any needed changes.

- This has to be plucked to make a sound. — banjo
- This has to be hit to make a sound. — triangle
- Air has to be blown into this to make a sound. — clarinet

This Is Why

Musical instruments use vibrations to make sounds. These sounds travel in waves. A percussion instrument produces a sound when its surface is struck. A stringed instrument produces a sound when its strings are vibrated. A wind instrument produces a sound when a column of air inside it vibrates.

Center Idea

Laminate the gameboard halves on pages 151 and 153 for durability. Tape the halves together and place them in a center along with a die and four game markers. Invite two to four players to visit the center and play the game as directed on the gameboard.

©The Mailbox® • Science in a Box • TEC60894

Musical Instrument Cards
Use with the activity on page 149.

guitar	banjo	cymbals	clarinet
gong	saxophone	harp	French horn
violin	flute	trumpet	xylophone
trombone	mandolin	tuba	snare drum
tambourine	cello	triangle	double bass
piano	bass drum	bongo drums	bagpipes

Answer Key: Percussion instruments: cymbals, gong, xylophone, snare drum, tambourine, triangle, bass drum, bongo drums. Stringed instruments: guitar, banjo, harp, violin, mandolin, cello, double bass, piano. Wind instruments: clarinet, saxophone, French horn, flute, trumpet, trombone, tuba, bagpipes.

MUSICAL

Board game spaces (in order): harp, bagpipes, bass drum, saxophone, gong, string bass, bongo drums, French horn, mandolin, tuba, tambourine, piano, Finish

Number Rolled	Type of Instrument
1 or 2	stringed
3 or 4	percussion
5 or 6	wind

151

FORCE AND MOTION

Objective: to understand that a force (a push or a pull) causes motion

Materials: page 155

Teacher preparation:
1. Follow the directions on page 3 to assemble the box.
2. Place this page inside the box. If desired, also store the remaining pages from this unit in the box.

Pull!

Activity

Begin by asking students to carefully watch what you do. Then perform a series of pushes and pulls using classroom objects. For example, pull the overhead screen down, push your chair under your desk, pull open the door, or push a pile of books across a table. Next, ask students to describe what you did. Explain that each movement was either a push or a pull that made something move and that this is called a *force*. A force changes the location of an object or it changes the direction in which an object is moving. Discuss each force that you performed and the resulting change in position. Then ask students to imagine what would happen if you had pushed the pile of books harder, or applied more force. Point out that the amount of force used determines how fast or how far an object moves. Divide your class into small groups and have each group make a simple chart like the one shown. Give groups about 15 minutes to fill their charts with as many examples of pushes and pulls as they can. When time is up, have each group share its list, allowing classmates to signal a thumbs-up or a thumbs-down to indicate the accuracy of each listed item.

Push	Pull
book cart	classroom door
ball on playground	drawer
a lawn mower	a toy train
the light switch	a sled
toy car	
baby stroller	

This Is Why

A force is a push or a pull that causes something to move. The amount of movement varies with the amount of force applied.

Center Idea

Laminate the cards on pages 157 and 159 and place them in a center along with the laminated center mat from page 156. A student decides whether each card shows a push or a pull and then places the card on the matching area of the center mat. Then he flips the cards to check his work.

©The Mailbox® • *Science in a Box* • TEC60894

Place each card on the matching wagon.

PUSH

PULL

156

Push Cards
Use with the center idea on page 155.

157

push	push	push
push	push	push
push	push	push

Pull Cards
Use with the center idea on page 155.

pull	pull	pull
pull	pull	pull
pull	pull	pull